SUSTAINABLE WATER SERVICES

A procedural guide

AF631453

SUSTAINABLE WATER SERVICES
A procedural guide

SWARD

Richard Ashley, Professor of Urban Water, University of Bradford
David Blackwood, Senior Lecturer, University of Abertay Dundee,
David Butler, Professor of Water Engineering, Imperial College London
Paul Jowitt, Professor of Civil Engineering Systems and Director of the Scottish Institute of Sustainable Technology, Heriot Watt University

Edited by *Nic Booker and Heidi Smith*

IWA Publishing

Published by IWA Publishing, Alliance House, 12 Caxton Street, London SW1H 0QS, UK
Telephone: +44 (0) 20 7654 5500; Fax: +44 (0) 20 7654 5555; Email: publications@iwap.co.uk
Web: www.iwapublishing.com

First published 2004
© 2004 IWA Publishing

Apart from any fair dealing for the purposes of research or private study, or criticism or review, as permitted under the UK Copyright, Designs and Patents Act (1998), no part of this publication may be reproduced, stored or transmitted in any form or by an means, without the prior permission in writing of the publisher, or, in the case of photographic reproduction, in accordance with the terms of licences issued by the Copyright Licensing Agency in the UK, or in accordance with the terms of licenses issued by the appropriate reproduction rights organization outside the UK. Enquiries concerning reproduction outside the terms stated here should be sent to IWA Publishing at the address printed above.

The publisher makes no representation, express or implied, with regard to the accuracy of the information contained in this book and cannot accept any legal responsibility or liability for errors or omissions that may be made.

Disclaimer
The information provided and the opinions given in this publication are not necessarily those of IWA or the editors, and should not be acted upon without independent consideration and professional advice. IWA and the editors will not accept responsibility for any loss or damage suffered by any person acting or refraining from acting upon any material contained in this publication.

British Library Cataloguing in Publication Data
A CIP catalogue record for this book is available from the British Library

Library of Congress Cataloging- in-Publication Data
A catalog record for this book is available from the Library of Congress

ISBN: 1 84339 065 5

CONTENTS

FOREWORD

Is it a curse 'to live in interesting times'? Certainly life is getting less straightforward, as we struggle with the challenges of sustainability, the Water Framework Directive, Climate change (flooding and drought), asset renewal, providing water services to new housing, not forgetting raising the funds to meet these objectives. This SWARD Guidebook has come to us at a good time, for the decisions that water utilities need to make are becoming more and more complex as they aim to balance often competing objectives.

In September 1997, the Engineering and Physical Sciences Research Council (EPSRC) launched the Water Infrastructure and Treatment Engineering (WITE) programme to encourage and promote research into novel and innovative engineering and physical science to help address the medium-and long-term strategic research priorities of the water utilities. The principal objectives of the programme were:

– To provide the basis for solutions to the problems faced by the water industry
– To promote research collaboration between universities and water utilities
– To encourage and promote research into the fundamental and strategic research issues relevant to the water utilities and related industries
– To promote multidisciplinary research networks so as to better harness the university skill base to tackle problems of national importance

SWARD was one of the projects funded by the WITE programme, and it has met all these objectives by bringing together engineers, scientists, economists and water utilities to develop a methodology to aid complex decision-making.

The UK water industry supplies 58 million people, has over 700,000km of water mains and sewers, more than 2,500 water treatment works and 9,000 sewage treatment works. The maintenance of this infrastructure and the provision of water and sewerage services requires long-term planning and investment. The total replacement value of these water industry assets is about £200 billion. Most of these assets are unseen, being buried under all our streets. In recent years spending on asset renewal has averaged a huge £1.5 billion each year but this is still a 133-year replacement cycle.

By investing at about £3 billion a year since 1990, the water utilities have delivered dramatic improvements in rivers, bathing waters, biodiversity and drinking water quality. But the assets that deliver these services to customers also need to be maintained. Based on the water companies' (England and Wales) preferred strategies (at December 2003) total capital investment between 2005 to 2010 would rise to around £21 billion, an increase of nearly 25% on the 2000 to 2005 period. Much of this increase is needed for asset renewal; for instance half the distribution system in central London is over 100 years old. But how do we choose which pipes to replace, in what order, with which methods?

The principal aim of SWARD is to help decision makers come to better decisions: better in terms of the consistency, openness and auditability of the decision process, and better in terms of the efforts made to appreciate and resolve (conflicting) objectives and preferences at the individual or organisational level or regulatory level. The techniques presented in this book are intended as additional decision-support processes, designed to complement the many already in use.

One measure of success of any project is whether it spawns further, related efforts. In that respect the SWARD initiative has led in part to the multi million pound EPSRC-funded project WaND (Water cycle management for New Developments), which we hope will provide another step forward by 2007.

By writing this book the authors have provided a service to all those involved in the complex world of water utility decision making. I commend it to you during these 'interesting times'.

Dr M S Farrimond
Director, UKWIR
London, February 2004

ACKNOWLEDGEMENTS

The authors would like to thank the following people that have contributed to the Sustainable Water industry Asset Resource Decisions (SWARD) project:

Daniel Gilmour, Sarah Hendry, Jim Moir and William Samson at the University of Abertay Dundee;
John Davies at Coventry University;
Hazem Gouda, Crina Oltean-Dumbrava and Heidi Smith at the University of Bradford;
Steven Cavill and Gary McIlkenny at Heriot-Watt University, Edinburgh;
Timothy Foxon, Matthew Leach and Peter Pearson at Imperial College London.

The authors would also like to express their gratitude to the Engineering and Physical Sciences Research Council and the UK Water Industry for funding the SWARD project.

SWARD

PREFACE

This Guidebook is the result of the Sustainable Water industry Asset Resource Decisions (SWARD) project, undertaken by a consortium of UK academics in collaboration with water service providers in Scotland, England and Romania. Its production follows consultation with a wide range of industrial and other stakeholding partners concerned with water services.

Why sustainability needs to be considered explicitly

Water service providers (WSPs) are faced with increasingly complex objectives and demands when making decisions related to asset management. This is because of the competing pressures to provide high quality services at low cost to the consumer whilst maintaining and protecting the environment. Following the Rio Earth Summit in 1992 and a number of associated activities, the need to ensure the sustainability of all development activity has been recognised and the key role of water service provision within this is reflected in recent initiatives, particularly the World Water Vision produced by the World Water Council. One of the major outcomes of the 2002 World Summit on Sustainable Development was a 'Framework for Action on Water and Sanitation' that outlines nine action areas. These include: to halve, by the year 2015, the proportion of people without access to safe drinking water and basic sanitation (reaffirmation of Millennium Development Goal); develop integrated water resources management (IWRM) frameworks, implement water management action plans at the country level; and mobilize financial resources to meet the investment needs in the water sector. The EC Water Framework Directive (CEC, 2000), which came into force in July 2000, takes a sustainable stance on many aspects of water management, and includes the requirement that member states adopt public participation (Article 14) in their approach to water management. The Directive will have to be implemented over the next 15 years. These wide-ranging initiatives have reinforced the need to define explicitly the relative weight to be given to the social, environmental and economic factors in decision-making processes. In turn, this has highlighted the difficulty for WSPs faced with choosing which options for managing their assets are likely to lead to better and more sustainable outcomes. It is necessary that the implications of sustainability in all other decisions that the WSP may make are considered, and in particular, the role of sustainability in its long-term strategic policy development. Explicitly incorporating a wider range of factors into their decision-making will be crucial if WSPs are to ensure that they play a full role in providing adequate water and sanitation world-wide and preserving and enhancing the quality of life in developed and developing countries.

The aim of this Guidebook

The aim of the Guidebook is to provide a framework that may be used by WSPs and associated stakeholders to enhance the way that sustainability is explicitly incorporated into decision-making processes related to water service provision. It focuses on the incorporation of sustainability criteria into investment decisions of the type associated with Asset Management Planning (AMP). The framework is open for all stakeholders to contribute their judgement of the relative importance of different sustainability criteria. A stakeholder is defined as any individual or group having a stake in how water and wastewater services are delivered, and could include customers, employees, regulators, environmental and community groups. Different stakeholders will assign different weights to criteria, and may well disagree as to the most sustainable option. By providing a common framework in which these differing opinions can be systematically investigated, the methodology presented within the Guidebook will facilitate discussion and increase the likelihood of arriving at a consensus position. The Guidebook is intended to complement rather than replace existing decision-making processes within the water industry.

The form of the Guidebook

The Guidebook consists of four main components. Firstly, sustainable development and how it relates to the water industry is discussed, leading to the identification of a range of relevant sustainability criteria. In the second part of the Guidebook decision making for sustainable development within the water industry is examined. An overview of the tools and techniques available for decision analysis, including multi-criteria decision-making procedures is also presented. The third part leads the reader through the seven phases of the SWARD DSPs. These are briefly outlined in Boxes i and ii. Although these are presented in a linear manner for ease of reference, it should be recognised that the DSPs will, in practice, be more iterative. The phases described may be ordered differently, or the user may step in and out of the DSPs according to: the scope and level of the project; nature of the objectives; or the approach to decision-making adopted within a particular organisation. Finally, the Guidebook includes two case studies that illustrate the application of the DSPs to real world examples. One of these case studies examines the management of sanitary waste disposed of via WC, and is presented in boxes throughout this Guidebook to help illustrate the SWARD decision support processes in more detail.

Feedback

The SWARD Guidebook is meant to be dynamic. The methodology presented here should be revised and updated as experience and knowledge is developed through its use. We invite feedback on the Guidebook; please send any comments/suggestions to:

The SWARD project, Pennine Water Group, School of Engineering, Design and Technology, University of Bradford, Bradford, West Yorkshire BD7 1DP, UK

Box i

THE SWARD DECISION SUPPORT PROCESSES I

Phase 1: Define objectives (Section 3.1): *This involves a review of past performance or the formulation of new targets, which leads to the setting of objectives for improving the performance of the water/wastewater system. A clear definition of the decision objective is the key to the subsequent DSPs.*

Phase 2: Generate options (Section 3.2): *Having defined the objectives it is necessary to generate a number of conceptualised options that will satisfy these objectives, one such option always being to do nothing. For stakeholders outside the WSP, it is important that they understand how the organisation makes decisions, by utilising a review phase.*

In most organisations, project development processes include value management workshops that are held at some stage of the development process. These workshops will generate alternatives, at least for the components of the system under development. In other organisations creative approaches may be more freely applied, potentially leading to a larger number of alternatives. The final stages of option generation must involve the reduction, by a process of elimination, of the options that are considered unsuitable. Ultimately a reduced number of options that appear to be of sufficient merit at this stage will be taken forward to the subsequent phases.

Phase 3: Select criteria and indicators (Section 3.3): *This involves the selection of specific criteria appropriate to the decision under consideration, from a set of generic criteria. This is necessary to enable evaluation of the relative sustainability of the proposed set of options, as some of the generic criteria will be irrelevant or impractical to apply in a particular situation. A list of primary, generic criteria has been generated. The sustainability criteria are categorised under the headings of economic, environmental and social, together with technical criteria, which relate primarily to the ability of the water/wastewater system to sustain and enhance the performance of the functions for which it is designed, taking due regard for the wider aspects of sustainability. Within each category, there are a small number of primary criteria. Under each of these primary criteria, a larger number of secondary criteria are specified. The master set developed in this Guidebook aims to serve a specific purpose - as an inclusive, generic set which can help to guide WSPs and their stakeholders in selecting a more specific set of criteria for use in comparing a range of options for meeting given objectives.*

Phase 4: Collect data and generate information (Section 3.4): *Phase 4 involves the collection or generation of data and information by which to score or rank each option according to the chosen criteria. Two broad types of approach for selecting a preferred option are: (1) comparisons using defined indicators; (2) formalised multi-criteria decision analysis (MCDA). Both types of DSP depend on the collection of indicator data and the generation of indicator information.*

Indicator data, in its raw form, may be suitable for direct use by stakeholders at the decision making stage. Alternatively, it may need to be processed into a more readily understood form (information generation). There are a variety of ways to collect data and/or generate information for relative sustainability assessment. Since future performance of the system under different options is being considered, data collection on past system performance will need to be combined with simulation modelling to assess the future performance of the option. There will be an interaction between the criteria chosen for assessment and the modelling tools used to estimate future performance of the system. The tools need to be appropriate to model the future values of the criteria under the range of options. A wide range of tools is available to collect data and/or generate information for relative sustainability assessment. These include: records or direct measurements; collecting data from stakeholders; modelling and/or simulation of existing and alternative systems; the use of economic analysis tools; and the use of environmental analysis tools.

Box ii

THE SWARD DECISION SUPPORT PROCESSES II

Phase 5: Analyse Options (Section 3.5): *There are many different tools, methods, techniques, algorithms, and systems that can aid and advise the decision-making process. The relative sustainability of water/wastewater systems should be assessed using the four main criteria categories: economic, social, environmental and technical, each of which are defined in terms of non-commensurate measures. This approach lends itself naturally to the use of one or more of the formalised multi-criteria decision-making methods as a screening tool.*

All decision support methods attempt to map the attributes of each decision outcome onto a preference structure via the various decision criteria. In some cases this results in a numeric function that expresses the perceived value or utility of each outcome compared with another (these are known as Aggregate Value/Utility Function Methods); in other cases this process produces a less precise ordering of the outcomes (known as Outranking Methods). In terms of input requirement, substantial work must be done to provide the necessary information needed for the analysis, whichever approach is used. For the process to be of value, it is important that any decision-making method contributes to the decision makers' understanding of the decision problem, rather than creating a focus of attention and distraction in its own right. In terms of output, any of the approaches can provide an overall ranking of the options under consideration, but this should be treated cautiously.

The most powerful approaches are those that provide a means of investigating the robustness of the data and the sensitivity of the overall ranking. The use of a transparent MCDA enables other stakeholders to criticise the weightings or choices used in the MCDA, and assess whether different values would affect the preferred choice of option.

Phase 6: Selection of the preferred option (Section 3.6): *MCDA tools provide information to guide the decision-maker about the relative sustainability of the options, but a further stage is required in the analysis before the final option selection. Analytical approaches do not remove the need for human judgement on issues such as risk and uncertainty and on the influence of intangible issues on the final decision. Examples of two risk analysis methods that can be used with the analytical tools to guide the decision-maker on the likelihood of reaching a wrong decision are sensitivity analysis and Monte-Carlo simulation. Sensitivity analysis involves making a change to the value of one variable in the analytical model, whilst retaining the original value of the other variables. In this way, the influence of that variable on the decision can be evaluated. The robustness of the decision can be tested under different scenarios by, for example, considering best and worst-case assumptions for key variables. Monte-Carlo simulation allows the effects of simultaneous changes in all the variables in the analysis to be observed.*

The decision-makers must consider each option in terms of the results of the decision analysis, the risk assessment of the robustness of this analysis and the qualitative information on intangible aspects. Ultimately they must use human judgement to make the final decision on which option is the more sustainable. The analytical process should overcome a number of cognitive limitations of the human decision maker, in that the number of pieces of information will have been reduced to manageable numbers and the risks associated with decision (in terms of relative sustainability) will have been identified.

Phase 7: Post project monitoring and feedback (Section 3.7): *Monitoring and evaluation should confirm that the evaluation undertaken in the decision stage was realistic; this is the decision review stage. It should also be used as process review as a means to refine the generic and specific criteria, the database and indicators (where used) and the SWARD processes. Post-implementation monitoring may be required over a significant period (possibly decades, depending on the nature of the development). For sustainable developments, certain aspects (particularly social indicators) will inevitably be 'monitored' informally. Follow-up monitoring for any project may illustrate that the criteria are robust and that the values are within the originally as-analysed ranges of sensitivity, in which case the selected option could be confirmed. It is also possible that the post-implementation phase reveals other criteria not originally included.*

GLOSSARY

Action a generic term used to designate the object of the decision or that which decision-aiding is directed towards. In practice the term action may be replaced by terms like: scenario, plan, project, proposal, variant, operation, investment, or solution, depending on the situation.

Actor any individual, group of individuals, entity or community likely to play any role directly or indirectly in the decision-making process. An actor is characterised by one value system (one preference structure).

AHP 'Analytical Hierarchical Procedure' A method of MCA in which the essential idea is to find the attribute trade-off weights through pairwise comparisons of attributes and, in addition to find the value of each decision alternative for each given attribute through pairwise comparisons of decision alternatives on that attribute.

Alternative a term used instead of action when two or more distinct potential actions can be implemented independently.

Analyst an actor who is responsible for the decision-aid process.

CBA 'Cost-Benefit Analysis' There are two types of CBA: Economic and Social. Economic analysis is used to assess the return on capital in an investment project over its economic life. It involves real cash flows that affect the investor. Social analysis measures the effect of an investment project on the community. Externalities are assessed in monetary terms and included in the cash flow forecasts.

CEA 'Cost-Effectiveness Analysis' A variant of CBA that assumes a common level of social benefit and therefore focuses on measuring the costs. This largely obviates the difficulty of estimating externalities.

CHAINET 'The European Network on Chain Analysis for Environmental Decision Support' This network links environmental stakeholders with experts on environmental analytical tools. The CHAINET book provides a toolbox for chain analysis and demonstrate the application of the toolbox to three case studies (Wrisberg *et al.*, 2002).

Criterion A measure against which options are assessed and a comparison is made of the extent to which they achieve stated objectives.

Decision-maker actor for whom the decision-aid tools are developed and implemented.

EFA 'Ecological Footprint Analysis' A sustainability assessment technique that uses the overall 'carrying capacity' of the planet as the basis of assessment. It links individual behaviour to organisational, regional and global targets using concepts such as the 'earthshare' – the average, sustainable, bioproductive capacity available per person. EFA is a purely environmental indicator.

ELECTRE 'Elimination et Choix Traduisant la Réalité' An outranking method where the predominant idea is that an alternative can be eliminated if it is dominated by other alternatives. The various versions of ELECTRE differ in their use of thresholds, weights, outranking relations and a credibility index.

Externalities Costs and benefits that are beyond the operation and interest of the providing authority or investor.

Indicators Tools that measure, simplify and communicate the level to which criterion are met. Indicators can be used to evaluate actions. For benchmarking progress, as post-project monitoring tools, in education for raising awareness of the issues involved in sustainable development. Indicators may be quantitative or qualitative.

Intangibles Externalities that defy monetary assessment.

LCA 'Life Cycle Assessment/Life Cycle Analysis' A comprehensive approach to analysing environmental burdens of a product, process or activity from 'cradle to grave'.

MAUT 'Multi-Attribute Utility Theory' A method of MCA based on the construction of a decision maker's utility function in order to represent his preference structure.

Monte Carlo Analysis A method whereby a distribution of possible outcomes is generated by repeated calculation of the values of the mathematical function (or functions) within a model. In each calculation, values for the constituent parameters are drawn by random sampling from predefined probability distributions.

MCA 'Multi-Criteria Analysis' Any structured approach to establish the overall preference between a set of options. MCA does not rely exclusively on monetary valuations, or the use of quantitative data.

NPV 'Net Present Value' The sum of the discounted values of future cash inflows and outflows.

Objectives The purpose that the water service provider wishes to achieve. A broad objective may be broken down into sub-objectives.

Option A structured way of achieving an objective.

Performance table/matrix a double-entry table showing actions in rows and criteria in columns. At the intersection of a row with a column we find the performance level of that action according to that criterion.

Preference Scale A scale representing relative strength of preference. The available options are used as anchors where the scale is relative. A score of 100 is usually associated with the most preferred option on a given criterion, and a score of 0 with the least preferred option.

PROMETHEE An outranking method, based on the ELECTRE procedure, which utilises the decision maker's preferences. The two versions of PROMETHEE differ by providing either a partial or total pre-ordering of the alternatives.

Qualitative Data data whose values are not numerical

Quantitative Data data whose values are numerical

'Quality of Life' The UK Governments sustainable development strategy launched in May 1999 aims to *'ensure a better quality of life for everyone now, and for generations to come'*. It sets out the key issues and priorities at a national scale. The strategy requires the meeting of four key objectives at the same time:

- social progress which recognises the needs of everyone
- effective protection of the environment
- prudent use of natural resources
- maintenance of high and stable levels of economic growth and employment

Reference Sustainability System (RSS) a methodology for representing the resource and material flows on which the environmental sustainability of a system depends. Enables an assessment of the potential of technologies and/or resource management strategies to enhance the sustainability of the system.

Scale set of elements ranked according to a complete order, reflecting the preferences of the decision-maker for an indicator.

Sensitivity analysis A process of varying one by one, in small increments, the values of the constituent parameters of a mathematical model to determine the effects of each of these changes on the outcome determined by the model.

SMART analysis 'Simple Multiple Attribute Rating Technique' Simple version of the MAUT procedure.

Stakeholders Any individual or group which has, consciously or not, a stake in how water and wastewater services are delivered. This could include customers, employees, environment and community groups. Future generations may also be stakeholders.

Sustainable Development There is a lack of a concise or universal definition for sustainable development. However, some general, and widely recognised, definitions are:

...development that meets the needs of current generations without compromising the ability of future generations to meet their needs and aspirations. (World Commission on Environment and Development, 1987)

...development that improves the quality of human life while living within the carrying capacity of supporting ecosystems. (International Union for the Conservation of Nature, 1991)

Sustainability Usually used as a shortened form of sustainable development; a guiding principle for efforts to reconcile environmental, economic and social imperatives.

Thresholds a concept that takes into account for each criterion the imprecision of certain data concerned with past or present phenomena.

Utility function A function which represents the decision makers preference or subjective valuation of the acceptability of an outcome, normally expressed as a scale of 0 (totally unacceptable, minimum satisfaction) to 1 (ideal solution, maximum satisfaction).

Value management A systematic, multi-disciplinary effort directed toward analysing the functions of projects for the purpose of achieving the best value at the lowest overall life cycle cost.

WSP 'Water Service Provider' A public, private or combined organisation that provides water supply and/or wastewater services.

SUSTAINABLE DEVELOPMENT AND THE WATER INDUSTRY

1.1 Sustainable Development

Sustainable development has become the key axiom of modern times, not just of committed environmentalists, but widely adopted across the political, business and social spectra. Why? Because it acknowledges that economic growth and environmental protection are not mutually incompatible, but can be pursued hand-in-hand. It is most closely associated with the UN's 1992 Earth Summit in Rio de Janeiro, but is perhaps most famously captured in Gro Harlem Brundtland's definition (WCED, 1987) as:

"Development that meets the needs of the present without compromising the ability of future generations to meet their own needs"

The roots of the Brundtland definition can be traced back at least to Jeremy Bentham's (1789) dictat that we should seek "the greatest good for the greatest number". And even the notion of intergenerational equity – so central to the Brundtland definition – was made explicit by American politician and forester Gifford Pinchot's (1947) addition of the phrase "for the longest time" to Bentham's imperative.

An alternative definition (IUCN/UNEP/WWF, 1991) asserts that sustainable development is that which "improves the quality of human life while living within the carrying capacity of supporting ecosystems". Here, the emphasis is placed on mankind's demand for and impact upon earth resources and the environment. Agenda 21 behoves us to "think global, but act local". Public participation becomes important and demands individual responsibility. The UK government is an enthusiastic proponent of sustainable development, with its own definition along the lines mentioned above (DETR, 1999). But it is the Brundtland definition that has endured, and has come to embody what in essence is a very simple idea: ensuring a better quality of life for everyone, now and into the future.

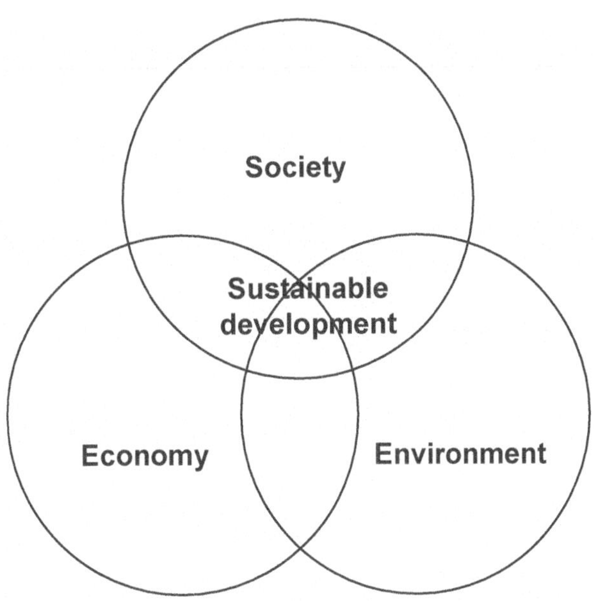

Figure 1.1: Sustainable development

In attempting to unpack the essence of sustainable development, three headings of utmost importance have emerged:

- Environmental protection
- Social inclusion
- Economic well-being

Sustainable development must be environmentally friendly, socially acceptable and financially viable (**Figure 1.1**). It is widely agreed that progress towards sustainable services requires the integration of these three elements into the decision making process.

Unfortunately the classification of the dimensions of sustainable development does not prescribe precisely how a particular set of policy or decision options are adjudged. Although the principle of sustainable development has gained widespread agreement, it could be argued that many organisations have still not grasped its meaning, and that even fewer have translated sustainability into action.

1.2 Sustainable Development and the Water Industry

One of the particular difficulties has been, and still is, that it is easier to define what is unsustainable practice rather than what is sustainable (Vachos & Braga, 2001). Nevertheless, specific water-relevant sustainability criteria have been proposed, against which projects may be judged, in the USA (ASCE/UNESCO, 1998) and UK (DETR, 1999), and elsewhere, and form a significant part of the development of this Guidebook. Progress has also been made in the water sector by a number of researchers (e.g. Butler & Parkinson, 1997; Larsen & Gujer, 1997; Niemczynowicz, 1994) who have proposed definitions of sustainable water management, and strategies to implement it.

The debate has also influenced legislation. The EC Water Framework Directive (CEC, 2000), which came into force in July 2000, takes a sustainable stance on many aspects of water management, including:

- Environment: achievement of good ecological status, no deterioration of water bodies into the future, elimination of hazardous substances at source
- Society: mandatory stakeholder participation in river basin plan production
- Economics: economic analysis of catchment-wide activity, recovery of costs for water services.

The Directive will have to be implemented over the next 15 years.

The UK Government has developed four sustainability principles (DETR, 1999) based on the various definitions outlined above:

- social progress which recognises the needs of everyone;
- effective protection of the environment;
- prudent use of natural resources;
- maintenance of high and stable levels of growth and employment

These sustainability principles are not the only ones that may be envisaged, but do provide an example of what principles should include. The objectives and principles could be applied to the water industry at two levels:

- At an overall corporate strategic level, for example in the WSP mission statement, and
- At an application level, where these strategic principles are being applied to a particular decision.

This SWARD Guidebook is designed to provide support principally for the second type of application; actually dealing with a particular development. However, the Guidebook will also inform the more strategic corporate level activity, both as a resource containing information about sustainability, and by feedback from application to inform the strategic processes of the WSP.

1.3 Sustainability Criteria

Sustainability criteria are defined here as the set of factors that may be used to assess which of a range of options, in this case for the development of water service provision, offers the greatest contribution to achieving sustainability objectives.

It is helpful, firstly, to draw a distinction between principles, criteria and indicators of sustainability. Principles are normative definitions or goals for sustainability. Criteria are the set of factors to be used to make a judgement about the relative sustainability of a set of options. Criteria may be evaluated using quantitative or qualitative information. Indicators can be used to measure the past and current values of specific criteria, and may be used to set standards against which future performance can be assessed. The relationship between principles, criteria and indicators is illustrated diagrammatically in **Figure 1.2**. The criteria and indicators are used to compare the set of options by measuring the attributes of the different options. Note that sustainability principles should remain constant over time, whereas the choice of criteria and indicators may change rapidly as knowledge advances.

Principles of sustainability aspire to a universal validity, which can be agreed upon by all. There have been a number of attempts to formulate such principles. One example is 'The Natural Step' (TNS) (Upham, 2000) which is based on four principles that are claimed to represent a scientific consensus on sustainability (Robert *et al.*, 1997). Other definitions are based on a single all-encompassing principle. Such principles, however, are necessarily abstract and idealised. In order to make judgements or decisions about which course of action to take, the guiding principles need to be made operational through the selection of more specific criteria and indicators.

The European Environment Agency (EEA) has proposed a classification system for indicators related to environmental systems. This uses a framework based on a conceptual 'Drivers-Pressures-State-Indicators-Response' (DPSIR) system, where the components are illustrated in **Figure 1.3**. Social, economic and technical developments (drivers) exert pressures on the environment, changing the state of the environment, impacting such that this ultimately elicits a response that feeds back to all stages in the system.

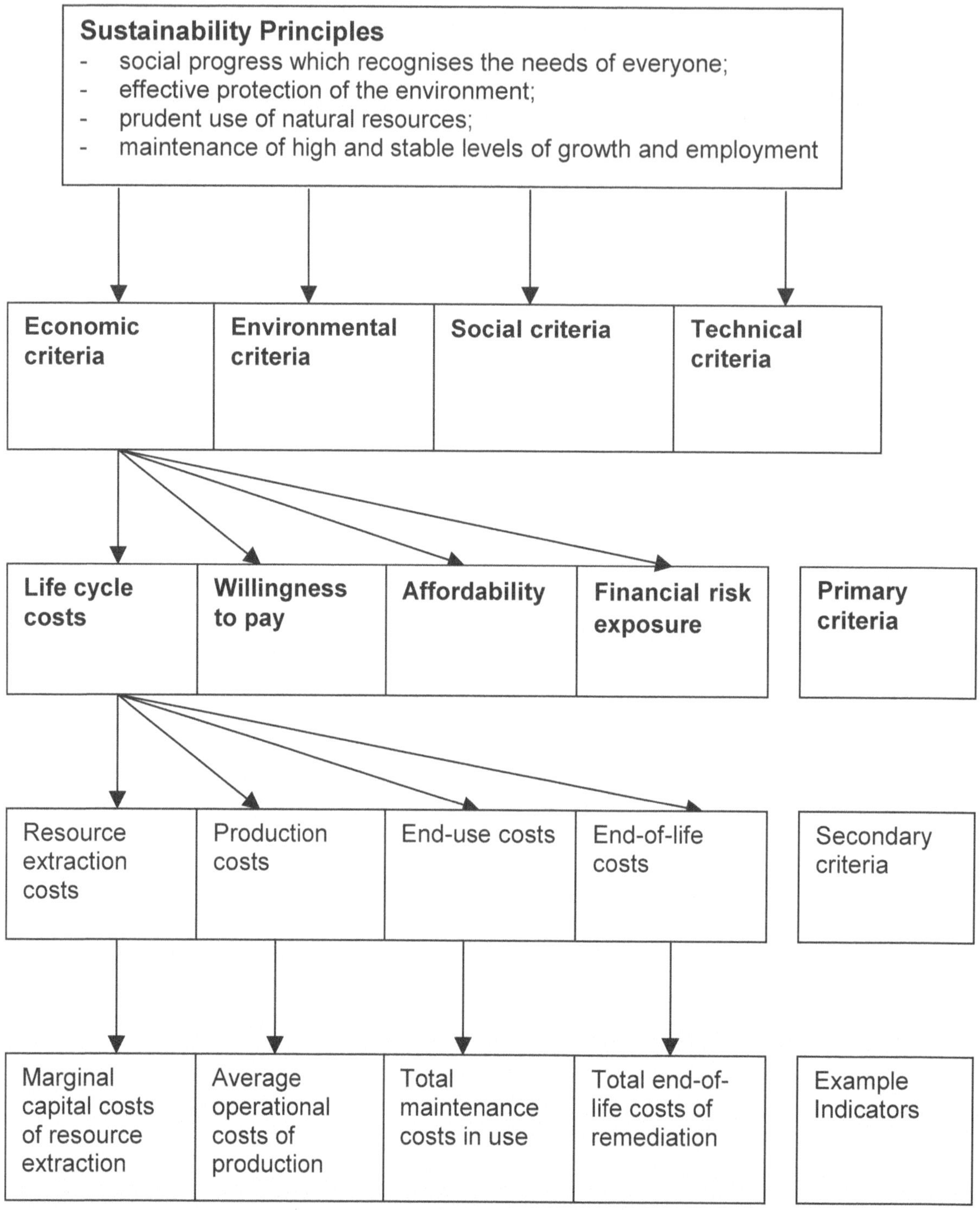

Figure 1.2: Relationship between sustainability principles, criteria and indicators (adapted from Foxon *et al.*, 2002).

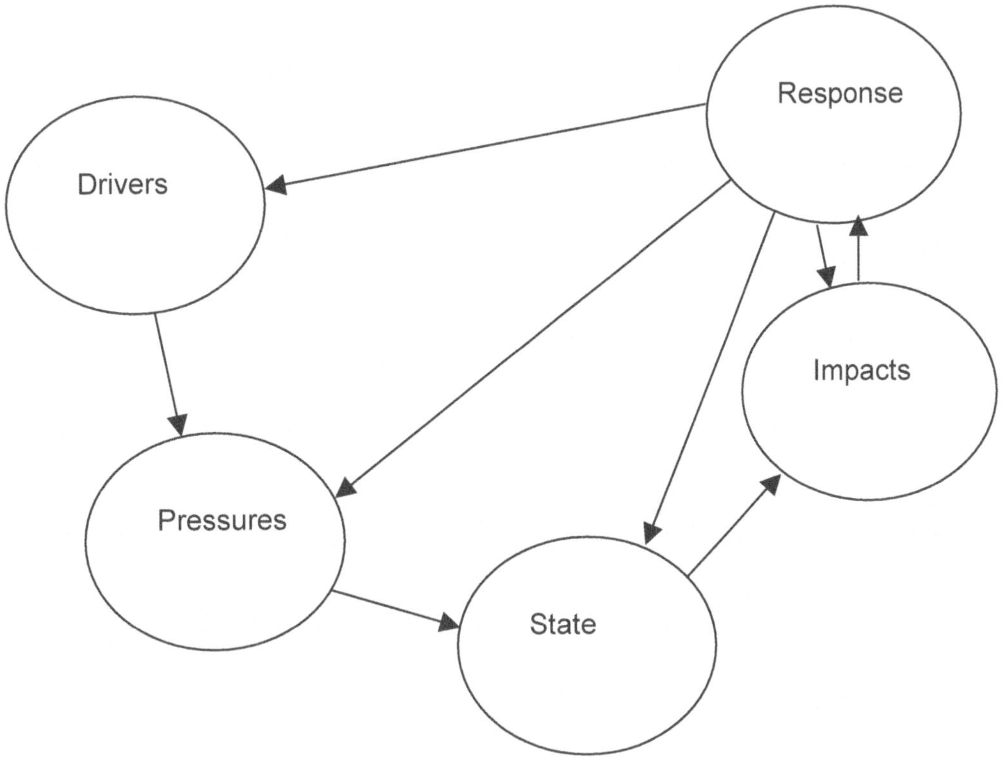

Figure 1.3: The DPSIR framework for reporting on environmental issues

1.3.1 International approaches to developing sustainability criteria

In trying to develop a generic set of criteria, it is essential to consider two key factors. What use will be made of this set of criteria? To what extent can any set of criteria encompass the range of issues to be considered under the heading of 'sustainability'? Some of these issues have been considered in approaches developed by other researchers.

One prominent and well-funded project on sustainable urban water management is being undertaken for the Swedish Foundation for Strategic Environmental Research (MISTRA). As part of this project, Jeppsson *et al.* (1998) examined the use of systems analysis for sustainable urban water management. Following a review of other approaches, they proposed a set of sustainability criteria and sub-criteria (Hellstrom *et al.*, 1999) under five categories: health and hygiene, socio-cultural, environmental, economic, and functional and technical. These were prioritised according to the relative impact from the water supply and sanitation system, compared to the total anthropogenic impact (Hellstrom *et al.*, 2000). They also concluded that, whereas several tools are available for analysis of environmental impact and resource utilisation, risk assessment and economic evaluation, methods for evaluating socio-cultural and functional criteria must be further developed.

A Dutch project (Balkema *et al.*, 1998; Balkema *et al.*, 2001; Balkema, 2003) constructed an overview of criteria used in 15 contemporary publications for the comparison of technologies with respect to sustainability. These criteria were grouped under the headings: economic, environmental, technical and social/cultural and included institutional factors. However, it was found that most publications did not incorporate social/cultural criteria directly into decision-making, but instead incorporated these indirectly into technical or economic criteria. Consequently, a set of sustainability criteria were proposed (Balkema, 2003) in which the demands of the end user are translated into functional criteria that have to be fulfilled by the technology. In order to perform its function, the technology uses resources from its environment and emits emissions into its environment. The resulting set of sustainability criteria is categorised into functional, economic, environmental, social and cultural.

A different approach, carried out in the context of the Dutch national research programme on Sustainable Technology Development (Graaf *et al.* 1997), identified a number of 'unsustainability factors' for urban-water systems. These were ranked in four levels of priority, from very important, such as emissions of heavy metals, to less important, including CO_2 emissions. They were used to assess a range of scenarios for reducing the unsustainability factors of the system by up to a factor of 20 by the year 2040.

Lundin *et al.* (1999) and Lundin (2003) produced a set of sustainability indicators covering environmental and technical factors and applicable to a wide range of different water and wastewater systems. The choice of indicators was guided by a number of principles: ability to demonstrate a move towards or away from sustainability; applicability to a broad range of systems; ability to provide an early warning of potential problems; availability of good-quality data; comprehensiveness of indicators; and cost-effectiveness. The indicators produced covered environmental issues and the efficiency and performance of the technical system. The variations in these indicators over time were used to assess progress of systems towards or away from sustainability.

One of the most comprehensive approaches to the issues and challenges raised by the concept of sustainability applied to water resource system design and management is the study produced by the International Hydrological Programme (IHP) together with the American Society of Civil Engineers (ASCE & UNESCO, 1998; Loucks & Gladwell, 1998). This report takes a holistic viewpoint and provides sustainability criteria and guidelines for water resource systems engineers and planners. The guidelines cover design, management and operation of the physical system; the environment and ecosystems; economics and finance; institutions and society; health and human welfare; and planning and technology. These lead to economic, ecological and environmental criteria for use in systems assessments of sustainable water resource planning, involving multiple stakeholders. The report recognises that "because sustainability is a function of various economic, environmental, ecological, social and physical goals and objectives, analyses must inevitably involve multi-objective tradeoffs in a multi-disciplinary and multi-participatory decision-making process."

The common thread running through these various approaches was thus the production of a structured classification of criteria, which would be forward-looking in order to be useful in decision making, and would be relevant to the decision-making process in question.

1.3.2 UK approaches to developing sustainability criteria

As discussed in **Section 1.2,** the UK government defines sustainable development as "ensuring a better quality of life for everyone, now and for generations to come" (DETR, 1999a). The principles that it sets out take the form of four objectives, to be met concurrently:

- social progress which recognises the needs of everyone;
- effective protection of the environment;
- prudent use of natural resources;
- maintenance of high and stable levels of growth and employment.

Based on these principles, a set of 15 headline indicators and 132 other core indicators (see **Appendix A**) were developed (DETR, 1999b). The trends in these indicators, either towards or away from meeting the relevant sustainability objective, will be reviewed annually. As the initial review (DETR, 2001) acknowledges, this type of indicator essentially looks backwards at the historical record, though attempts have been made to include forward projections towards future government targets. Moreover, these indicators tend to look at the cumulative effects or results of actions, such as 'days when air pollution is moderate or higher' or 'percentage of river network assessed as being of good or fair quality'. Such indicators serve a very useful purpose in assessing the overall results of many individual decisions by a number of stakeholders, and so in guiding the development of policies to influence those decisions in the direction of increasing sustainability. However, they are less useful in aiding an individual decision-maker, such as a WSP, to decide which of a potential range of options would offer the greatest contribution to sustainability.

Water UK, the trade body for all UK WSPs, developed a similar set of 25 indicators (see **Appendix A**) for use by the UK water industry to measure its progress towards environmental sustainability (Water UK, 2000). These indicators serve a useful purpose by presenting an aggregate national picture of trends towards or away from sustainability, but they are again less useful in guiding future decision making. Furthermore, these initial indicators only covered environmental sustainability, and not social or economic aspects. A new set of indicators, which include social and economic aspects, are now out in draft form.

There appears to be no means of avoiding the need to make value judgements in the definition of sustainability and hence criteria. Therefore any process working toward sustainable development must make explicit the judgements involved.

For these reasons, it was believed necessary to develop a generic set of sustainability criteria for use in this Guidebook, which would be informed by the UK Government and Water UK principles and indicators, but which would be of direct use in assessing development options. These sustainability criteria have been developed in collaboration with the water industry and stakeholders, using information gained from the focus groups and decision mapping of current decision making processes in the UK water industry, together with reviews of sets of criteria which have been proposed for similar purposes, mostly in other European countries.

1.4 Water Industry Appropriate Sustainability Criteria

The approach taken here aims to draw on the strengths of the various approaches discussed above to develop a structure appropriate to UK WSPs that would encapsulate the richness and diversity of sustainability. This is based on a hierarchy, consisting of a small number of high-level primary criteria, which would cascade down to a larger number of more specific secondary criteria. These criteria are classified under four categories:

- economic
- environmental
- social
- technical.

These categories aim to encapsulate the economic, environmental and social principles of sustainability, together with technical criteria, which relate primarily to the ability of the water/wastewater system to sustain and enhance the performance of the functions for which it is designed.

Within each category, a small number of primary criteria are specified (see **Table 1.1**). Under each of these primary criteria, a larger number of secondary criteria are specified (see **Tables 1.2a to d**). These can be used to formulate indicators to assess the future performance of the water/wastewater system under the particular development option under scrutiny. These may then be used to assess whether the system is moving towards or away from sustainability. The primary criteria are given in **Table 1.1**, and example indicators are given in **Table 1.2a to d**.

Table 1.1 Set of Primary Criteria

Category	Primary criteria
Economic	- Life cycle costs - Willingness to pay - Affordability - Financial risk exposure
Environmental	- Resource utilisation - Service provision - Environmental impact
Social	- Impact on risks to human health - Acceptability to stakeholders - Participation and responsibility - Public awareness and understanding - Social inclusion
Technical	- Performance of the system - Reliability - Durability - Flexibility and adaptability

1.4.1 Economic criteria

Life cycle costs cover all stages of the life cycle from resource extraction, production, end-use and end-of-life, including capital, operation and maintenance components, and marginal, average and total cost categories (e.g. Penning-Rowsell *et al.*, 2003; Skipworth *et al.*, 2002). The second criterion covers willingness to pay for attributes covering product, environmental, safety and health factors. This can be evaluated directly for those attributes that are traded and priced, but, for other attributes, the willingness to pay has to be deduced indirectly, for example from contingent valuation studies or from the price of similar commodities. Affordability relates to the ability of householders to pay for services delivered. Financial risk exposure relates to the risk of loss to the company associated with particular kinds of investment.

1.4.2 Environmental criteria

Resource utilisation covers use of water resources, land, energy, materials and chemicals. Service provision relates to the provision of water and wastewater services by the WSP. Environmental impact covers a range of impacts on water, land, air and bio-diversity (e.g. Matos *et al.*, 2003).

1.4.3 Social criteria

Impact on risks to human health includes availability of clean water, risk of infection and exposure to toxic compounds. Acceptability to stakeholders covers acceptability and perceptions of impacts on health and environment. Participation and responsibility includes individual actions, participation in decision making and willingness to change behaviour. Public awareness and understanding covers awareness of implications of behaviour and consumer information. Social inclusion covers poverty alleviation, voluntary activity and access to water courses. Sustainability engages here in terms of promoting a sense of environmental awareness and the notion that we cannot obtain something for nothing, which in turn impacts upon how people regard and use water as a resource. In this respect then, water is not simply a material resource but is also inter-twined and inseparable from people's lifestyles and social attitudes (e.g. Strang, 2001).

1.4.4 Technical criteria

Performance of the system includes the quality of supplied water and treated wastewater. Reliability of the system includes the availability and distribution of water supplied and wastewater, water usage restrictions and pressure of mains water, and system failure in providing these services, including flooding and pollution from sewers. Durability of system covers the expected lifetime of the system. Flexibility and adaptability covers the ability to make future changes to the system (e.g. Matos *et al.*, 2003).

While the set of criteria is intended to encompass the broad range of sustainability, as with any set of criteria, the selection and categorisation are to a certain extent arbitrary. This set aims to serve a specific purpose - as an inclusive, generic set which can help to guide WSPs and their stakeholders in selecting a more specific set of criteria for use in comparing a range of options for meeting given objectives. For this reason, the primary criteria cover broad types of economic, environmental, social and technical impacts that a given option may have. The secondary criteria give more specific types of impact under these headings. The generic criteria were reviewed throughout the development of the SWARD decision support processes, though use of workshops held with WSP project partners in the UK and in Romania.

Table 1.2a: Primary and Secondary Economic Criteria and Example Indicators

Primary Criteria	Secondary Criteria	Example Indicator (units)
Life cycle costs *Stages:* • Resource extraction costs • Production costs • End-use costs • Remediation costs	Capital Costs Operational Costs Maintenance Costs Decommissioning Costs	Average costs in £ £/year £/year Average costs in £
Willingness to pay	… for the product … for environmental benefits … for safety … for health … for other attributes	£/unit of service delivered £/unit of benefit £/unit of reduced risk £/unit of reduced risk
Affordability	% of household budget for lowest income households	% of budget
Financial risk exposure	… for capital investment … for other investments	Risk of loss associated with investments

Table 1.2b: Primary and Secondary Environmental Criteria and Example Indicators

Primary Criteria	Secondary Criteria	Example Indicator (Units)
Resource utilisation	Water resource use	
	- Withdrawal	Annual freshwater withdrawal/ annual available volume, %
	- River water quality	% of rivers of good or fair quality
	- Nutrients in water	% of river length with greater than guideline nutrient concentrations
	Land use	Land area used in km2
	Energy use	
	- Energy for water supply	Energy use (kWh/cu. m)
	- Energy for wastewater treatment	Energy use (kWh/cu. m)
	Chemical use	
	- WTP or on-site (herbicides)	Litres/yr
	Material use	
	- aggregates, plastics, metals	Total material requirement (Tonnes/yr)
Service provision	Water services	
	- water consumption (household)	Litres/capita
	- water consumption (non-household)	Ml/year
	- leakage rates	Ml/year
	- Water re-use	% water reused
	Wastewater services	
	Wastewater production	Ml/day
Environmental impact	Impact on water	Loads of BOD, P and N
	Impact on land	
	- Sludge re-use	Tonnes of sludge re-used
	- Recovery of nutrients	P and N recovered/ total incoming P and N
	- Quality of sludge	Heavy metal content
	Impact on air	
	- CO_2, SO_2, NO_x emissions	Tonnes/year
	Impact on biological diversity	No. of key species at risk

Table 1.2c: Primary and Secondary Social Criteria and Example Indicators

Primary Criteria	Secondary Criteria	Example Indicator (Units)
Impact on risks to human health	Availability of clean water	% of population with access to clean water
	Risk of infection	Number of affected persons /100,000 population
	Exposure to toxic compounds	Number of affected persons /100,000 population
Acceptability to stakeholders	Acceptability to stakeholders	Number of complaints /100,000 population
	Perceived health and safety impact	% of 'users' with concerns about injury, drowning, risk of infection
	Perceived environmental impact	% of 'users' perceiving a positive environmental impact, e.g. reduced flood risk, habitat creation, amenity benefit
Participation and responsibility	Participation in sustainable behaviour	Number of people participating in initiative /catchment population
	Individual action	Number of Local Agenda 21 meetings
	Willingness to change behaviour	Number of people willing to change behaviour /catchment population
Public understanding and awareness	Awareness of implications of behaviour	% awareness in local community
	Stakeholder information	Information made available to customers and other stakeholders
Social inclusion	Social inclusion	% of population with access to information
	Voluntary activity	% of population involved in voluntary activity
	Community spirit	Number of local community groups
	Access to watercourse	%of population with access to watercourse

Table 1.2d: Primary and Secondary Technical Criteria and Example Indicators

Primary Criteria	Secondary Criteria	Example Indicator (Units)
Performance	Water quality (supply)	
	- Quality of supplied water	Compliance with required standards in tests performed throughout the year (%)
	- Water quality complaints	No. of water quality complaints per year (e.g. aesthetics)
	Water quality (treatment)	
	- Compliance with consent conditions	Compliance with consent conditions (%)
	- Effluent quality	Quality of effluent
Reliability	Water availability and distribution (water supply)	
	- Raw water availability (DG1, OFWAT)	Population whose calculated water resource availability is below the reference level (%)
	- Water use restrictions (DG4)	Population who have been subject to water usage restrictions (%)
	- Restriction or interruption complaints	Number of restriction or interruption complaints in a year
	- Mains water pressure (DG2)	Number of customer's properties that are at risk of receiving mains water pressure below reference level
	- Pressure complaints	Number of pressure complaints in a year
	System failure (wastewater)	
	- Flooding from sewers (DG5)	Number of customer's properties where risk of flooding is greater than the reference level
	- Risk of failure	Risk of failure to meet consent conditions due to treatment process malfunction (qualitative)
Durability	Design life	Number of years system expected to operate successfully for
Flexibility and adaptability	Flexibility of the system	Level of accommodation in design: potential and ability to accommodate future changes (Qualitative)
	Ability to add to or remove from system	Cost (£) of adding or removing from system in response to future changes

DECISION MAKING FOR SUSTAINABLE DEVELOPMENT

2.1 Drivers for improving the quality of Water Industry decisions

Globally there is a drive to improve the performance of WSPs (e.g. Matos *et al.*, 2003). In developed countries this means more effective and efficient services for maintaining public health and utility. In the developing world the need is for basic service provision and whilst sustainability considerations are important to ensure longer-term provision, they may be considered as something of a luxury until there is equity of access to safe water supplies and sanitation. Nonetheless the WSPs around the world are now operating in a global economy and the way in which they make decisions is fundamental to the attainment of more sustainable water systems (Wrisberg *et al.*, 2002). Whether the WSPs are private companies, publicly owned, or operate in some other institutional mode, their adoption of transparent and stakeholder sensitive decision making processes will be crucial for future changes to water service provision. A major difficulty in the explicit inclusion of sustainability in the decision support process, is accounting for the various contributory elements: social, technical, economic and environmental.

Historically, some aspects of sustainability have been central to the development of water resources. This is particularly true in respect of providing clean water supplies and safe methods of wastewater treatment and disposal to the environment. The objectives were to enhance both the social and economic well-being of the community. In many cases, these activities also led to environmental improvement. In the UK the Water Act 2003 reinforces the need to take better regard for sustainability in the way in which water services are delivered and regulated.

So why the present concern about the need to embed sustainability criteria more formally within the water industry? Simply because the scale of human activity has made serious – and unsustainable – impacts on the environment. Not only in terms of water use and water quality, or even in the impacts on other natural and biological resources, but also with respect to its impact on social and economic activity. The effective and efficient provision of water services is central to economic activity and quality of life in an increasingly competitive world. WSPs therefore have to balance the needs of society, economy and the environment, and in so doing, make decisions on infrastructure development that are socially useful, economically efficient, environmentally responsible and which can be sustained over time.

The need for better support for decision-making processes is identified in a number of initiatives and was confirmed by the decision mapping exercises undertaken in the development of this Guidebook (Bouchart *et al.*, 2002). Many of these initiatives have been directed toward the need to provide information pertaining to environmental systems and impacts.

In the UK, the Foresight programme operates via a number of panels and task forces to assist Government with forward thinking about key issues. The Environment Appraisal Task Force has addressed the problem of environmental/sustainability appraisal in two documents 'Making Sustainability Count' and 'Towards more sustainable decisions'. Environmental appraisal and sustainability appraisal are presented as being synonymous in the latter and four general conclusions were taken to apply across all stakeholder sectors:

- The environmental, social and economic strands of appraisal must be brought together to inform decisions. These may need to be assessed separately, but must be reported together for the decision process.

- Social appraisal techniques need further development and should take better account of the distribution of impacts on different sections of the community
- Appraisal has to move up the decision-making ladder to the strategic levels of policies, plans and programmes. To the extent that the responsibility for conducting appraisals moves to officials rather than professionals in the assessment community, there will need to be much more emphasis on simple approaches backed up by suitable training and awareness.
- Evidence-based policy making means that there will need to be more emphasis on monitoring and evaluation throughout and after the life of a project or programme. Indicators and targets will need to be built in right from the beginning.

The facilitation and encouragement of greater public involvement in appraisal is advocated for all of the principal *actors* in the decision support process. The EC Water Framework Directive specifically requires member states to adopt public participation (Article 14) in water management. In October 2003 the EC also adopted a package of three legislative proposals to align Community legislation with the requirements of the 1998 Aarhus Convention. The legislative package consists of three proposals:

- Proposal for a DIRECTIVE OF THE EUROPEAN PARLIAMENT AND OF THE COUNCIL on access to justice in environmental matters;
- Proposal for a REGULATION OF THE EUROPEAN PARLIAMENT AND OF THE COUNCIL on the application of the provisions of the Aarhus Convention on Access to information, Public Participation in Decision-making and Access to Justice in Environmental Matters to EC institutions and bodies;
- Proposal for a COUNCIL DECISION on the conclusion, on behalf of the European Community, of the Convention on access to information, public participation in decision making and access to justice regarding environmental matters.

In response to recent reports by the House of Commons Environmental Audit Committee and the Environment Sub-Committee, the UK Government has included in the Water Act 2003 the duty to give the new Water Services Regulatory Authority (replacing OFWAT) a clearer role in promoting more sustainable management of water, in recognition of the fact that in the water industry, the interaction of environmental and economic issues is particularly strong. In addition, the CORE (Corporate Responsibility) Coalition was launched in 2002, along with the publication of its draft Private Member's Bill. The Bill demands (amongst other things) that companies should consult with their stakeholders on company activities and impacts; directors should take all reasonable steps to minimise any negative social, environmental and economic effects of their companies' operations; and stakeholders would be able to hold companies and directors legally to account on these issues. Water UK, in association with Ernst and Young, has responded with a 'source' documant setting out the industry's perspective on global sustainability.

The water industry employs much higher than average levels of financial capital per unit of turnover than most other sectors, typically 12% compared with the electricity industry (4%), gas industry (2%) and supermarkets (0.4%). Much of the capital investment is bound up in long term physical assets, many of them of considerable age and many of them buried (Whelan, 2003). The need for WSPs to use their infrastructure and asset base to meet their business objectives in the most efficient and sustainable way possible is therefore of enormous importance. Asset Management Planning (AMP) – coupled with innovation and change in operational practice - has become the established way of achieving this, not simply as part of the regulatory process, but increasingly to reflect a much sharper focus on customer needs and wider stakeholder expectations. The Water Services Regulatory Authority promotes improvements in ideas about how best this may be done and a lot of information about past AMP processes in England and Wales and future thinking is given on the Regulator's web site. It is timely to utilise this guidebook as the Water Act in England

and Wales is set to place a specific sustainability duty on the Director General of OFWAT, and the successor, the Water Services Regulatory Authority, which will influence the way in which water company economic plans are developed in future. In addition, the sustainability dimension has recently been reinforced as a key component of the operation of the Environment Agency. This is amply illustrated by a number of reports on the Environment Agency website. In Scotland, the Water Industry (Scotland) Act 2002 has recently given Scottish Water a general duty to act in the way best calculated to contribute to the achievement of sustainable development. This duty must inform all its decisions, including those on funding new infrastructure.

A key element of AMP involves examining various Key Performance Indicators (KPI's) associated with customer service, compliance and other corporate objectives/drivers, and then relating these to the condition, performance and criticality of the assets. This exercise should not simply be retrospective but should be forward looking. Recently a number of proposals for performance indicators for the water industry have been developed. For example, a manual from the International Water Association (IWA) for water supply services suggests an approach that may be applied worldwide (Alegre *et al.*, 2000) which is now being extended to wastewater services (Matos *et al.*, 2003). The objectives of this approach include: creating a framework of indicators that may allow for future comparisons between water undertakings in the scope of metric benchmarking initiatives, and from which sub-sets of indicators can be extracted, according to specific need, for the use of regulatory agencies. As an example, the system set out in the water supply document incorporates six groups of indicators: water resources, personnel, physical, operational, quality of service, and financial. Of the complete group of 133 indicators, 26 have been rated as top priority. In the wastewater manual the processes of planning and selecting the KPIs is extended, compared with the earlier manual.

One of the major challenges facing water service providers relates to capital maintenance. This is the repair, refurbishment, or replacement of each asset, or alternative provision, as the asset deteriorates with age and use. Maintenance may be either reactive in response to a failure, or proactive in order to prevent asset failure. Recently, in England and Wales, the water service providers have invested in the development of a 'Common Framework' for capital maintenance (Heywood *et al.*, 2002) that provides an agreed framework to estimating future capital maintenance that is economic. The approach is risk-based and goes someway toward a full sustainability approach, although it does not fully include social and environmental criteria.

Some of the water industry's key drivers, such as regulatory and statutory compliance, financial performance etc, are explicit. Others, such as customer confidence and reputation, are less so. All of them influence the objectives a WSP sets for itself and the decisions it takes to achieve them.

The Common Framework for Capital Maintenance Planning, developed building upon the concepts outlined in the Ofwat letter MD161, is based on the analysis of risk and encompasses an economic approach which allows the trade-off between capital and operational cost options to be considered. The framework enables an economic level of capital maintenance to be identified with due regard to the costs associated with asset failure and (where appropriate) the value placed by customers on service improvements (UKWIR, 2002). However, the results of a recent survey commissioned by OFWAT to investigate customers' views and needs in respect of the water industry (MORI, 2002) found that respondents exhibit a relatively low awareness and concern for their water and sewerage services. There was also a lack of connection made between the level of water and sewerage bills and the source of any money needed to fund improvements. This highlights the need for caution when integrating any stakeholder-specific issues into decisions about asset management investments.

The operation of most WSPs has moved on from a position where the only concern was financial capital and how to maximise the return on it. Other forms of capital are to be invested and nurtured to bring about sustainable returns that can benefit the business in the longer term. For example, the Sustainability Assessment Model (SAM) being tested by BP deals with economic, resource, environmental and social capital of the oil and gas industry. The water industry is no exception, and in many ways, the nature of the water service business lends itself well to this more holistic approach. These other forms of water industry capital are those related to the environment (natural capital), infrastructure, and people (staff; customers and communities). There is no longer any KPI, but rather a set of KPI's operating at different levels in the business and appropriate to particular performance criteria, be they economic, environmental, technical, social or whatever.

A major element of the SWARD framework is therefore concerned with extending the ideas of AMP so that it favours making sustainable choices in the process of asset management and investment rather than those that are superficially efficient at the simple technical or financial level. This should conform better with the requirements of recent legislation such as the EC Water Framework Directive and the UK Water Act.

All of this indicates the complexity of implementation of the simple concept described above. Yet, in practice, decisions have to be made and, increasingly, justified in wider terms than just cost effectiveness.

This Guidebook is intended to show how appropriate sustainability criteria can be incorporated by the Water Service Providers into their decision-making processes.

2.2 Types of decision supported

Many recent publications on decision making make reference to early work by Simon (1976), March and Simon (1958) and Lindblom (1959), who all noted that decision making in practice is seldom structured and that often satisfactory solutions are reached in an ad-hoc way. Astley *et al* (1982), reviewing the above, concluded that "decision making can be seen as a process of muddling through towards a satisfactory and sufficient outcome as opposed to necessarily obtaining the optimal solution". March and Simon question the ability of the rational decision-maker to make optimal choices and distinguish between optimal and satisfactory solutions to problems.

They describe a solution as being optimal if:

- there exists a set of criteria that permits all known alternatives to be compared;
- the alternative in question is preferred, by these criteria, to all other alternatives.

An alternative is satisfactory if:

- there exists a set of criteria that describes minimally satisfactory alternatives;
- the alternative in question meets or exceeds all these criteria.

They conclude that most human decision-making is concerned with the discovery and selection of satisfactory rather than optimal alternatives and describe this process as "satisficing". The SWARD procedure is intended to provide both a means of identification of relevant criteria and also methodologies for the assessment of each of these criteria. However, it should be appreciated that the nature of the data required to support the application of SWARD will vary depending on the stage in the overall evolution of the solution at which the decision is required. Cross (1989) suggested that these design related problems are particularly difficult to deal with because they are ill-defined or ill-structured, therefore, the highly specified and clearly defined environment which is necessary for rational decision making does not exist. He summarises the characteristics of an ill-defined problem as follows.

- There is no definitive formulation of the problem. The goals are vague, many constraints and criteria are unknown and the problem is complex.
- Any problem formulation may embody inconsistencies. Conflicts and inconsistencies will exist in the problem and will emerge in the process of problem solving.
- Formulations of the problem are solution dependent. The way in which the solution is conceived influences the way in which the problem is conceived.
- Proposing solutions is a way of understanding the problem. Constraints and criteria emerge as a result of evaluating a solution.
- There is no definitive solution to the problem. Optimality is difficult to define and different solutions may be valid responses to the original problem (again consider Simon's concept of satisficing).

Simon (1984) discussed the relationship between well-structured and ill-structured problems. Well-structured problems, he argued, are at the other extreme of a continuum of degrees of definiteness from ill-structured problems and he suggested criteria for well-structured problems that are broadly opposite to those of Cross above. He further stated that:

> *"In general terms the problems presented to problem solvers by the world are best described as ISP's (Ill-structured problems). They become WSP's (Well- structured problems) only in the process of being prepared for the problem-solvers. It is not exaggerating much to say that there are no WSP's, only ISP's that have been formalised for problem solvers"*

Rodgers defines non-programmed decisions as those which are ill-structured, necessary on a non-recurring basis and requiring substantial analysis on behalf of the decision maker. Programmed decisions are well-structured, are necessary on a recurring basis and algorithms or other structured knowledge exists to support the decision maker. Rodgers typifies such decisions as design decisions using codes of practice. Other authors refer to the higher level decisions as strategic and the lower level decisions as operational, but the characteristics of the decisions are similar.

In addition consideration must be given to identifying the decision-maker. Some alternatives are:

- Individual: a stand-alone decision-maker, decision can be influenced by the individuals cognitive limitations and personal experience, values, perception and intuition.
- Multiple: a varied number of stakeholders with different experience, values, perceptions and beliefs.
- Group: a number of people with broadly similar perceptions and beliefs, each with equal authority.
- Team: a number of people with broadly similar perceptions and beliefs, but with levels of authority being largely dictated by the organisation hierarchy.

For many water companies and authorities a team approach to decision making is used. During decision making at higher levels of these organisations, a move to the use of a multiple approach to decision making may lead to improved inclusion of sustainability issues into decision making.

Figure 2.1 illustrates the context and demonstrates the complexity of issues and stakeholders that need to be incorporated in decision making within the water industry so that sustainability issues are included.

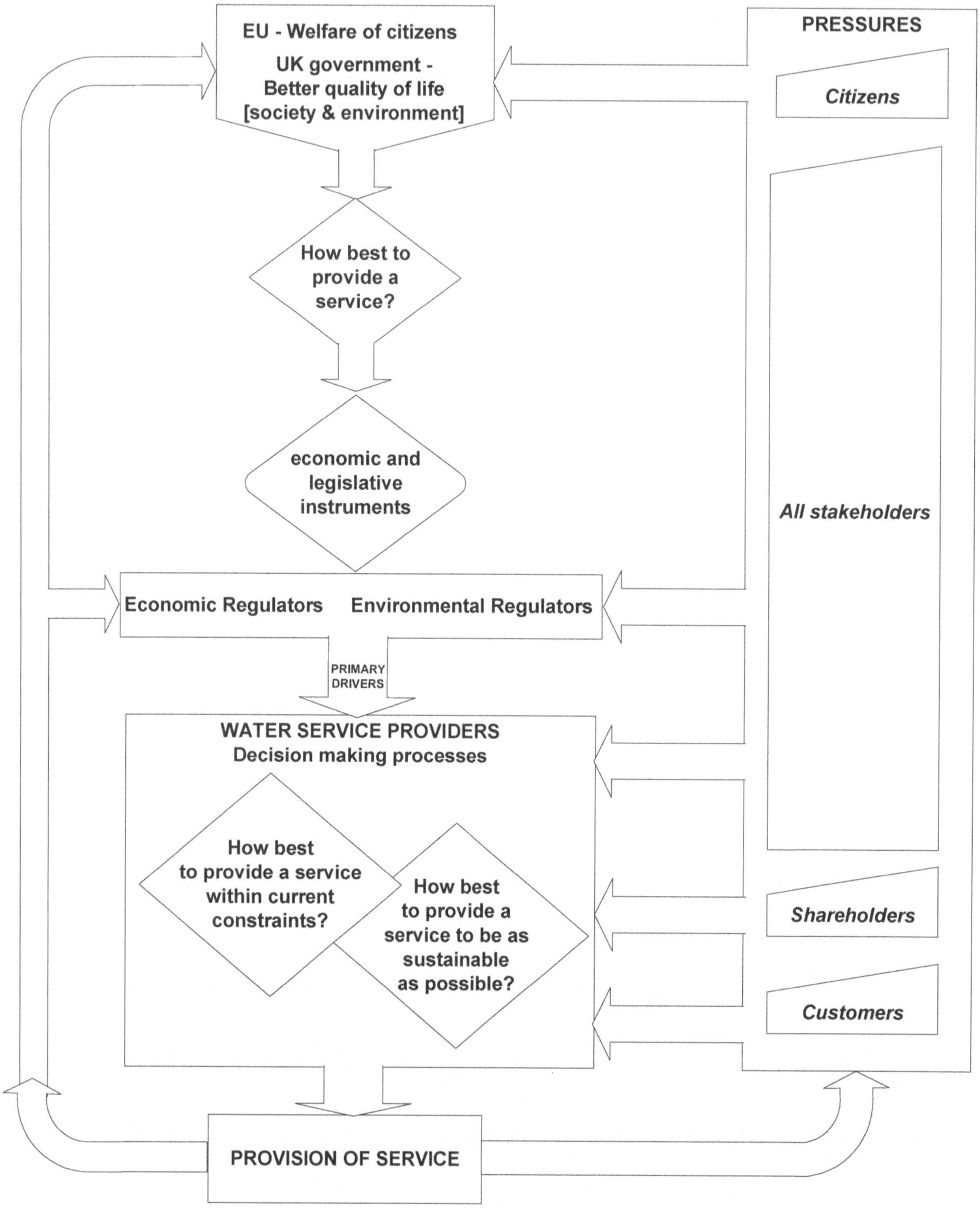

Figure 2.1: The decision-making context

2.3 The SWARD concept

The principal objective of the SWARD Guidebook is, firstly, to facilitate the explicit inclusion of sustainability in the decision-making processes for water service asset planning and, secondly, to provide the means whereby the relative sustainability of each of the options under investigation can be compared. It is not prescriptive, and it is accepted that there may be compelling circumstances that lead to the actual selected option being less than ideal from the sustainability perspective. The critical issue is to ensure that decisions are reached through informed choice so that the sustainability implications of the preferred option are clearly identified. For example, legislative changes (regional or national) may introduce the need to adopt investment options that are not as sustainable as they may be, due for example, to the need to implement asset changes to meet very short time scales. In these circumstances the value of the Guidebook is not simply to determine which solutions are most sustainable, but to highlight where policy requirements are likely to lead to solutions that are less than ideal.

Water Service Providers (WSPs) make decisions at a range of scales – in magnitude, space and time and at different strategic levels. The Guidebook has been developed and applied to example case studies for which some or all of these factors are significant, and where the WSP decision processes both inform and are informed by the drivers and pressures for change. In the former case, the explicit consideration of sustainability criteria will help suggest the need for new drivers, i.e. in Europe, new directives (or changes to existing directives) that will drive the implementation of solutions that are more sustainable. In the latter case, sustainability issues will arise from decisions of the EU (e.g. the requirements of EU-wide environmental legislation such as the Water Framework Directive) or the UK Government (e.g. competition policy and economic regulation). Guidance such as that provided here may also be utilised to help inform future legislation. For example, Anglian Water have recently challenged the implementation of certain aspects of the Urban Wastewater Treatment Directive on the grounds that it would lead to solutions that would be likely to be less sustainable. This was demonstrated using an approach based on the ecological footprint of the proposed development needs required to comply (Mylius, 2002).

During the SWARD project a range of decision examples were investigated (Ashley *et al.*, 2002), these are presented in Table 2.1. Two of the case studies undertaken as part of the SWARD project are presented in Appendix D of this Guidebook.

Table 2.1: Types of decision exemplified within the Guidebook

Generic Examples of Decision Type	Specific Examples of Decision Type
a. How can an existing service be provided in a better way? (Can include new constraints, legislation and technology)	· More sustainable management of domestic sanitary waste (see Appendix D2)
b. Has a new service to be provided? (i.e. existing service inadequate – new development or new constraints)	· Provision of First Time Sewerage to unconnected properties
c. Asset Management Planning/Infrastructure Development	· Water Distribution Network management/rehabilitation (see Appendix D1)
d. Choices between different ways of providing a service (alternative treatment processes etc.)	· Sustainable Urban Drainage Systems
e. Choices as to where a new service should be sited (here, there or anywhere)	· Options/location for wastewater treatment facilities

Note that the above are not mutually exclusive, merely examples of relevant types of decision

Figure 2.2 shows the overall framework and processes described in this Guidebook. These comprise seven phases. The phases are described in **Sections 3.1 to 3.7** of the Guidebook. The phases of the decision support processes are:

1. Definition of decision objectives.
2. Generation of options.
3. Selection of criteria and indicators relevant to the decision in question and the options under consideration – some will be qualitative and others quantitative.
4. The collection of data and generation of information for each option and each criterion, together with risk evaluation results. This will entail analysis of each option in terms of each specific criterion – utilising appropriate analytical tools.
5. Analysis of options – weightings and ranking of results, where appropriate, and risk assessment/sensitivity analysis.
6. Selection and implementation of preferred option(s) using some form of multi-criteria assessment.
7. Post project monitoring and feedback to inform future decision-making.

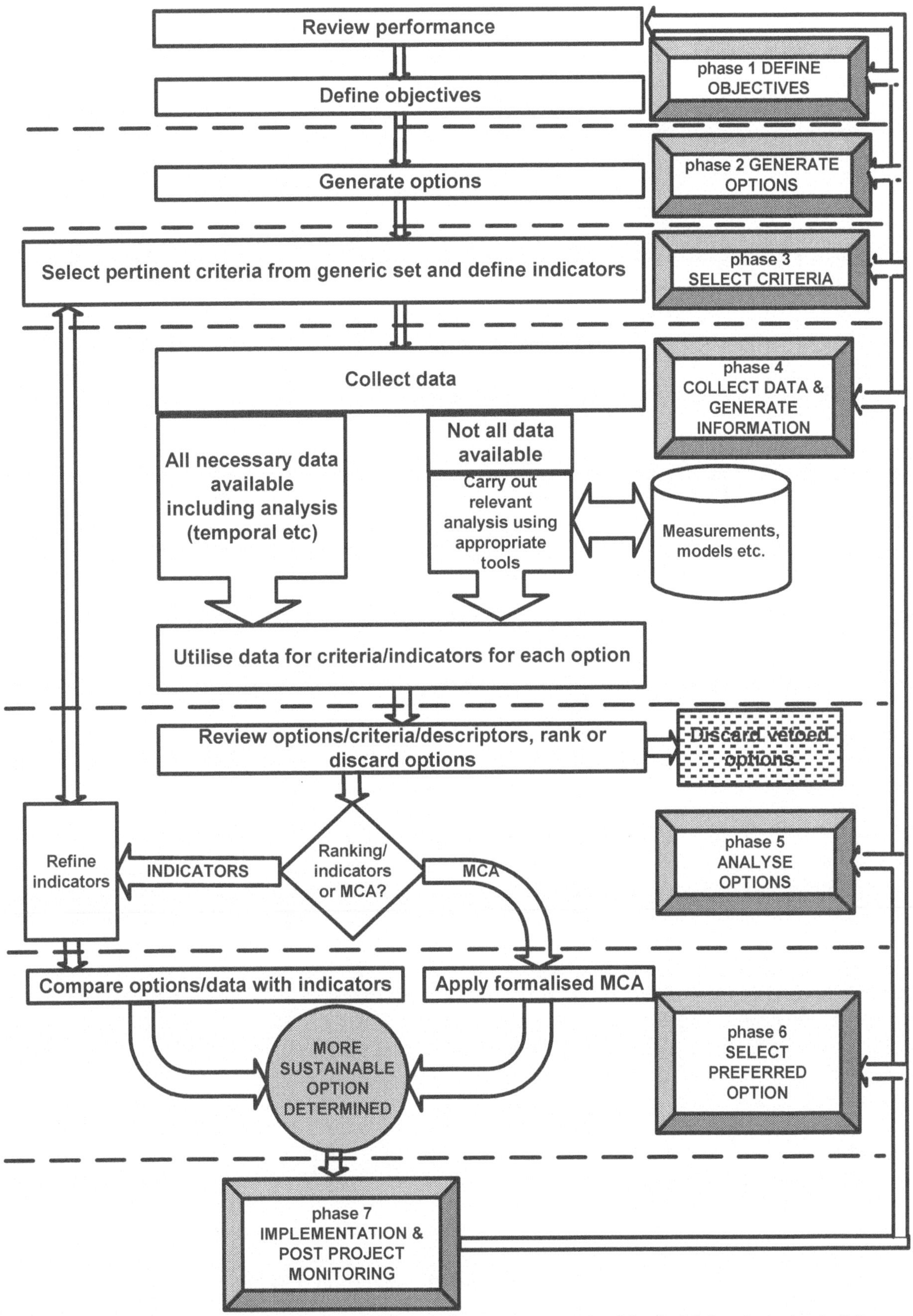

Figure 2.2: Phases in the decision making processes used in this Guidebook (adapted from Foxon *et al*., 2002)

2.4 Options appraisal and decision making

The SWARD Guidebook is primarily concerned with making asset management and related decisions in the water industry. Decision-making is not an exact science and the associated terminology in the literature can vary. It is therefore appropriate that the key elements of a decision are introduced and some of the associated decision-making terminology, as used in this guidebook, is briefly explained and defined.

At one level, the elements of a decision are straightforward: they consist of choices (i.e. the design or decision options) and the exercise of preferences over those choices, in the face of a variety of constraints. What makes many real life decisions more difficult, such as those that are encountered in the water industry, is in discriminating between the complexity of the choices available and defining the decision-maker's preferences over them. This is further exacerbated where there is no single decision-maker (or where the decision maker(s) are acting in some form of proxy for others). These situations are referred to as Multi-Stakeholder or Multi-Decision Maker Decision Problems.

Decision-making can be further complicated by the presence of uncertainty, where the exercise of a particular choice does not guarantee the actual decision outcome. For example, any decision to increase the provision of water supply capacity against uncertain estimates of consumer demand and the unknown effects of climate change cannot determine precisely the consequent level of supply reliability or the operating cost. Decisions such as this therefore include an element of risk.

Thus all decisions have four elements: the choice of action, the influence of uncertainty on the eventual outcome, preferences over those outcomes and constraints that limit the feasible choices.

If there is no uncertainty, then the decision reduces simply to finding the choice that leads to the preferred outcome, while satisfying the constraints. If there is no choice, there is no decision, and if there is no preference, then any feasible choice will do.

Although this Guidebook examines some of the risk issues, it is directed mainly at the problem of establishing preferences and making decisions in situations where the uncertainties are either small or where they are taken into account through the selection of appropriate worst case scenarios (for example by using the suitable design storm flow, the 25 year drought yields or through some form of event-based or random statistical simulation). It is acknowledged that this is not ideal, but as will become apparent, the more pressing and difficult problem is the determination of the decision-makers' preference structure over the possible outcomes. So for the most part, uncertainty will be assumed absent, and so any choice of a particular design or decision option will be supposed to determine the dominant outcome. The focus will be on establishing the underlying preference structure.

Associated with each design/decision option will be a number of characteristics or attributes. For example, for a conventional sewerage system these might include Capital Cost, Pipeline Diameter; Maintenance Cost; Hydraulic Capacity; Life Cycle/Whole Life Cost; Pipe Material; Expected Lifetime; Failure Frequency; Repair Cost; Traffic Disruption; Health and Safety Requirements etc. Each design option is therefore said to be Multi-Attributed.

A single issue such as cost might dominate the decision-makers' preferences over the various design options, but more generally preferences will reflect the satisfaction of a number of criteria. The selection of the preferred design option can therefore said to be a Multi-Criteria Decision Making (MCDM) Problem. The term Multi-Objective Decision Making (MODM) problem is also sometimes used to reflect the same idea but is used by some to imply a degree of optimisation in the search for the best solution rather than the selection of best solution from a set of pre-defined alternatives.

In terms of this Guidebook, the focus is on choosing between design options in terms of their overall sustainability. For this four high level criteria categories have been suggested: economic, environmental, social and technical. The first three categories represent the well-known triple bottom line of sustainability, whilst the technical criteria relate to the ability of the water/wastewater infrastructure to sustain and enhance the performance of the functions for which it is designed and operated.

Some of the attributes, either individually or in some combination might be interpreted as (directed) indicators of one of these decision criteria. For example a desire to achieve high economic efficiency will somehow be reflected by issues such as Capital Cost, Maintenance Cost; Life Cycle/Whole Life Cost; Expected Lifetime; Failure Frequency; etc. In a different combination - and sometimes even in a different directional sense – the design attributes might be taken to reflect a different decision criterion, such as the need to reduce social impacts. In this case, the appropriate combination of attributes might be Life Cycle/Whole Life Cost; Pipe Material; Expected Lifetime; Failure Frequency; Traffic Disruption; Health and Safety Requirements etc. And even within this social criterion there might be conflict between meeting customer expectations in terms of levels of service and ability to pay.

Decisions in the water industry are generally both Multi-Attributed and Multi-Criteria. It is useful to remember that the Attributes are simply non-judgemental objective characteristics of the design options; whereas the Criteria are a reflection of the preferences of those making the decision.

The question naturally arises as to how the attributes should be measured and combined to provide an indication of how well each criterion is satisfied and then how the relative degrees of satisfaction of the various criteria are brought together ("weighted") to arrive at the most preferred design option. One approach is to provide a numerical measure of satisfaction for each criterion and then weight the criteria in some way to arrive at an overall quantitative evaluation of each design option. A number of formalised MCDM methods using this type of approach are described in this Guidebook and illustrated in one of the case studies, and for some problems this is an acceptable method.

Unfortunately formalised methods of MCDM have two main drawbacks, which ironically are at their worst when the decision problem is complex and potentially more controversial:

- The apparent objectivity can imply an unwarranted degree of precision; and
- The process by which the measures and weights are determined is both difficult conceptually and, where there are several decision-makers, can be extremely contentious.

In such situations, a more informal approach is preferred, in which the decision is arrived at by informed discussion of the relative merits of each decision option, its attributes and the interpretation of these attributes against the (jointly agreed) decision criteria. What is often most effective is to combine both approaches – using the formalised MCDM approach as a screening tool and then refining the short list of preferred decisions by an open and auditable discussion/analysis. This approach is also described in the case studies.

2.5 Tools to assist decision making

As explained above, the very nature of such decision problems is that they are usually complex: the decision options are often multi-attributed; there are usually several stakeholders/decision-makers, each one seeking to satisfy several decision criteria according to their own (or their perception of their organisation's) preference structure. In order to assist in the evaluation of the range of economic, social, environmental and technical criteria, a number of decision-making approaches and tools can be applied. This section describes how a number of the more important decision-making approaches and tools can be applied in the context of assessing the sustainability of water/wastewater systems.

The techniques presented here are intended as additional decision-support processes, designed to complement the many already in use by the WSPs. There are many different tools, methods, techniques, algorithms, and systems that can aid/advise the decision-making process. It is, however, essential that a distinction be drawn between procedural tools and analytical tools. Wrisberg *et al.* (2002) note that procedural tools 'focus on procedures to guide the way to reach a decision' whereas analytical tools provide 'information as to the consequences of a choice'.

Within Wrisberg's definition, the SWARD framework as given in the guidebook can be best described as a procedural tool, within which other procedural/analytical tools may be used.

2.5.1 Decision Analysis Methods

The principal aim of multi-criteria decision analysis is to help decision makers come to better decisions: - better in terms of the consistency, openness and auditability of the decision process, and better in terms of the efforts made to appreciate and resolve (conflicting) objectives and preferences at the individual or organisational level. The quality of the decisions and their realisation depends on the quality of the decision-making and implementation processes. To be effective these processes should contain a strong element of learning: learning more about the problem, learning more about one's own and other stakeholder preferences, and learning more about organisational values and objectives.

The decision process should contribute to:

- Establishing the decision-making context by identifying the actors, the various possibilities of action, their consequences, the stakes, etc.
- The organisation and/or structuring of how the decision-making process unfolds in order to increase coherence between the values underlying the perceived goals and those embodied in the final decision.
- Getting the actors to cooperate by providing a decision framework favourable to mutual understanding and debate.
- The elaboration of recommendations using results taken from models and computational procedures conceived within the framework of a working hypothesis.
- Maximising participation in the final decision and securing "buy in" from all the stakeholders.

2.5.2 Limits to Objectivity in Decision-Making

The multi-criteria and multi-stakeholder nature of decision making in complex problems has emphasised their subjective nature. Complex decisions cannot be reduced to a simple mechanistic process and there are real limits to objectivity. Some of the most common reasons why such objectivity is difficult to achieve are as follows:

- The borderline between what is and what is not feasible is often fuzzy and frequently modified in the light of what is found through the analysis itself.
- In many real world problems, several actors/stakeholders take part in the decision process and there is often a distinction (and sometimes some confusion) as to who is the final arbiter. Is the decision being taken collectively, or does it need to have the support of the key decision-maker, or is the decision really just a recommendation to some other body?
- The decision-makers' collective and individual preferences are seldom well shaped and include areas of uncertainty. Even where individuals can define their own

preference structures fairly crisply, there is no unique and objective way of combining individual preferences into a group preference structure.

- The data values or the probability distributions affecting some of the decision variables are in many cases imprecise and/or defined in an arbitrary way;
- In general, it is impossible to say that a decision is a good one or a bad one by referring only to a mathematical model. The organisation, robustness and culture of the whole decision process that leads to a given decision also contribute to its quality and success.

Nevertheless, a number of methodologies have been developed that attempt to cast decision making into a rational and more mechanistic framework, and whilst these should not be used to resolve the decision problem alone, they can be very useful as a screening tool and in illuminating some of the key issues and conflicts, given the particular difficulties posed by the breadth of the sustainability issue.

2.5.3 Multi Criteria Decision Making Methods

Essentially all decision support methods attempt to map the attributes of each decision outcome onto a preference structure via the various decision criteria. In some cases this results in a numeric function that expresses the perceived value or utility of each outcome compared with another (Aggregate Value/Utility Function Methods); in other cases it produces a less precise ordering of the outcomes (Outranking Methods) (DETR, 2000).

The difficulty is generally one of trying to compare the value of different entities expressed in different units (or sometimes with no units at all, but simply using linguistic descriptions of the alternatives). In the simplest case, a numeric utility function might be linear (i.e. such that a decrease in one attribute can be traded off linearly for a gain in another). Generally, the tradeoffs are not linear; for example, at the extremes where a surfeit of one attribute does not compensate for the dearth of another; or where attributes have synergistic benefits or dis-benefits. Eliciting this utility function across the full range of the underlying attributes is not a trivial task for even a single decision-maker even if facilitated by a decision analysis expert and supported by decision analysis software. There is a danger that the focus becomes the theory behind the decision model and not the decision itself.

Aggregate Value/Utility Function Methods

The three most prominent groups of aggregate value/utility function methods are:

1. *Multi-Attribute Utility Theory* (MAUT) – based on the construction of a utility function representing the decision maker's preference structure. Typically, the first step in assembling the utility function will be to establish how the decision criteria are related to the attributes of each decision outcome. These relationships might be linear or non-linear. The next step is to determine how the criteria might be combined to give an overall measure of value/utility. Again, this might be linear or non-linear. If it is linear in the criteria, then overall utility will simply be a weighted linear sum of the performance against each sustainability criterion (Keeney and Raifa, 1993).
2. *Simple Multiple Attribute Rating Technique* (SMART) – a simplified version of the MAUT procedure above (Edwards and Newman 1982).
3. *Analytic Hierarchy Process* (AHP) – the essential idea here is to find the attribute trade-off weights through comparison of pairs (pairwise) of attributes and find the value of each decision alternative for each given attribute through these pairwise comparisons of the decision alternatives (Saaty, 1980).

Outranking Methods

This approach differs from the value/utility function methods in that it rejects the notion that all alternatives are comparable, maintaining that in some circumstances a decision maker will be unwilling or unable to compare some options. The two most prominent groups of outranking methods are:

1. ELECTRE – the central idea is that an alternative can be eliminated if it is dominated by other alternatives (Roy, 1991). The various versions of ELECTRE differ in their use of thresholds, weights, outranking relations and credibility indexes.
2. PROMETHEE – based on the ELECTRE procedure and which utilises the decision-maker's preferences. The two versions of PROMETHEE differ by providing either a partial or total ordering of the alternatives (Brans *et al.*, 1984).

There are other schools of thought that are less prominent in the field of multi-criteria analysis and not so well documented in the literature. These include:

1. Multi-Dimensional Scaling Methods - used in conjunction with other analyses as a means of identifying clusters of similar sets of data points through visual representation.
2. Methods based on Fuzzy Sets – used in conjunction with both aggregate value and outranking approaches. The major contribution of these methods is to emphasise that judgements are not precise and to seek to incorporate this in the analysis in an explicit way.
3. Ideal Point Methods - the underlying concept is to compare the performance of each option by reference to the ideal outcome and measure its "distance" from it in some way.

In terms of input requirement, substantial work must be done to provide the necessary information needed for the analysis, whichever approach is used. For the process to be of value, it is important that any decision-making method contributes to the decision makers' understanding of the decision problem, rather than creating a focus of attention and distraction in its own right.

In terms of output, any of the above approaches can provide an overall ranking but this should always be treated cautiously. The most powerful approaches are those that provide the means of investigation of the robustness of the data and the sensitivity of the overall ranking.

2.5.4 Use of Multi-Criteria Decision-Making Methods within SWARD

The relative sustainability of water/wastewater systems within the SWARD Framework is assessed using four main criteria:

- Economic
- Social
- Environmental
- Technical

This approach lends itself naturally to the use of one or more of the formalised MCDM methods as a screening tool. More detail and examples of five of the most common MCDM methods is given in **Appendix C.**

2.5.5 Value management

Value Management is described as:

"A systematic, multi-disciplinary effort directed toward analysing the functions of projects for the purpose of achieving the best value at the lowest overall life cycle cost." (Norton & McElligott, 1995).

A Value Management study normally comprises three main phases: pre-workshop; workshop; and post-workshop, which are described briefly below and in more detail in **Appendix C**.

Pre-workshop phase
This phase is concerned mainly with workshop organisation and provision of information to participants (stakeholders). The client and Value Specialist will normally meet to establish study logistics, project issues and constraints. The Value Specialist will compile the relevant information (client brief, drawings, surveys, costs, etc) that is sent to participants prior to the workshop.

Workshop phase
Workshops typically last for one to six days, depending on the project. Workshops follow a 'job-plan' consisting of the following phases:

- Information phase:
  - Overview of VM process by Value Specialist.
  - Presentations by design team.
  - Functional analysis of project.
- Creative phase:
  - Generate alternative ideas to achieve same basic function.
  - Rationale may be lower costs/improved functions/both.
  - Brainstorming technique often used.
- Evaluation phase:
  - Ideas from creative phase evaluated, best ideas selected.
  - Economic and non-economic evaluation criteria used.
  - Various evaluation methods may be employed.
- Development phase:
  - Selected ideas developed into fully detailed proposals.
  - Proposals include descriptions, pros/cons, initial and life-cycle costs, technical calculations, sketches, etc.
- Presentation phase:
  - Proposal summaries presented to decision makers.
  - Draft copies of summaries handed over to decision makers. Action plan written detailing follow-up work.

Post-workshop phase
A workshop report, containing a clear and comprehensive record of events and outcomes, is prepared and circulated. Decision-makers are given an appropriate period to consider the recommendations of the VM team before an implementation meeting is held. The purpose of this meeting is to determine whether proposals should be accepted or rejected, and to establish subsequent actions.

VM Evaluation Methods
This is likely to be the most problematic of the various phases within Value Management. This is when the workshop team must somehow arrive at a consensus on which proposals to reject and which to select for further development. A variety of evaluation methods are typically used in this phase. Some of these will be straightforward economic/life cycle or technical evaluations. Others will attempt to capture the multi-criteria aspects of the problem and these might include some of the techniques referred to above, or related techniques that also involve ranking and comparing critical issues. Norton & McElligott (1995) have described the more common "evaluation techniques" under three categories:

- Selection by Value Management Team Co-ordinator (VMTC)
- Simple democratic
- Complex democratic

The most relevant of these techniques, within the above categories are described in **Appendix C** of this Guidebook, along with some of the criticisms of Value Management. The difficulties of arriving at "democratic" decisions within value management are no less than those encountered generally in multi-stakeholder decision problems and referred to earlier. This is why it is better to regard formalised MCDA as a screening tool, or a way of exploring candidate solutions, rather than as an objective surrogate for the decision-maker.

2.5.6 Use of Value Management within SWARD

Value Management can be easily and usefully accommodated in the SWARD Framework, and may help improve the inclusion of sustainability in the following ways:

- it is multi-disciplinary, so a variety of stakeholders can be included;
- analysing function to achieve best value (do what is actually needed) is in keeping with sustainability principles;
- the functions of projects can include social and environmental functions;
- an emphasis on life cycle cost is better, in terms of sustainability, than focussing purely on initial cost.

The case study presented in **Appendix D1** illustrates a decision-making procedure that was used by a WSP during the development of a high-level water resources and distribution strategy. Within this, VM was used as the main procedural tool and exhibits some features in common with the processes set out in this Guidebook. The steps used within this VM approach were broadly consistent with the SWARD DSPs although not necessarily conducted in the same order as these are presented in this Guidebook. It is important to stress that the SWARD Guidebook processes are to be interpreted in that sense – as a guide. The context of a particular application, and the other exigencies that inevitably occur within a WSP will inevitably mean that the decision process has to be adapted to meet local and temporal circumstances and the influence of other ongoing developments with the business of the WSP.

DECISION SUPPORT PROCESSES – IMPLEMENTING SWARD

3.1 Phase 1 - Define Objectives

Phase 1 of the SWARD framework is about Reviewing Performance, Determining Priorities and Defining Objectives. It takes place against a broad background, and where the internal stakeholders might have differing perceptions of the priorities for the business, its customers and its regulators, and hold different perceptions of the value and contributions of its sources of capital.

It is expected that WSP objectives will already be formulated within the context of the discussion in **Section 1** and be continually subject to dynamic revision as circumstances change. As far as the Guidebook is concerned, the relevant objectives are those that impinge upon sustainability considerations. In a truly integrated perspective, this would include all objectives and activities of the WSP. However, the way in which sustainability is currently considered is normally explicit, as a separate concept, in much the same way as environmental considerations used to be dealt with. Ultimately, sustainability should become an intrinsic part of every aspect of a WSPs business, and hence the organisation would not then require explicit guidance on how to include sustainability into its decision-making processes.

Users of this Guidebook should initially consider the place of the over-riding SWARD objective in their organisation at the corporate level and the application level, improving the relative sustainability of the way in which they provide services. **Box 3.1** presents the decision objective for the sanitary waste case study. The evolution of this case study is illustrated within the boxes found throughout this chapter.

At the application level, sub-objectives can be specified to achieve the over-riding objective above. These must include the four high level groups of criteria: Economic, Environmental, Technical and Social if sustainability is to be given due consideration. Subsequent application of the SWARD DSPs may show that one or more of these high level groups is not significant to the particular decision under consideration, but each of these must be included at the initial stage of objective formulation.

Users of the Guidebook should explicitly define a set of principles that elucidate how they intend to interpret the economic, environmental, social and technical objectives of sustainability in the context of their operations or their personal perspectives. These principles will provide high-level guidance and direction in the decision support processes, and should be generally applicable to the range of WSP activities. The principles should broadly remain constant over time.

Box 3.1

CASE STUDY

Management of domestic sanitary waste disposed of via WC

PHASE 1 - DEFINE OBJECTIVES

The presence of SW in the environment has a damaging effect on the amenity value of beaches in the town and the surrounding area. The decision for a problem relating to SW escape may be prompted by legislation or may be customer driven, i.e. a high number of customer complaints about the number of items of SW on beaches.

The decision objective in relation to this case study is:

> ***'to manage the number of sanitary waste items escaping to the environment via storm overflow discharges in the catchment in a way that is likely to be the most sustainable'.***

3.2 Phase 2 – Generate options

Having defined the objectives it is necessary to generate a number of conceptualised options that will satisfy these objectives, often one such option is to do nothing. It is important to appreciate what approaches are used for design and decision making processes. This is because there are a range of contrasting views on the nature of option generation. At one extreme, option generation is viewed as being purely creative and at the other it is viewed as a mechanistic, systematic process (Cross, 1994, Starkey, 1992). Of course, in most decision making processes, the decision maker(s) are unaware of where their approach fits in this spectrum of types of process.

The classical mechanistic approach to concept generation consists of a systematic analysis of the problem. The following stages may be involved in the mechanistic type of option generation:

- Abstraction of the most important elements comprising the problem to be addressed
- Identifying the requirements of each of these elements and determining the relationships between these requirements
- Combining these relationships into the possible solutions. These solutions are known as "concept variants".

It is recognised that the identification of the concept variants will often involve intuition in which the solution comes to the decision-maker in a flash after a period of search and reflection. Other approaches place more emphasis on the intuitive or creative nature of option generation. A distinction can be made between this and the development of ideas where there is a ceiling. The ceiling is a constraint influencing the decision, and may be for example, a technological boundary, or even a conceptual boundary.

Approaches to concept generation include:

- Pragmatic Design - the use of existing available materials and methods without innovation.
- Lateral Thinking - the use of ideas from other fields applied to the problem.
- Brainstorming - the generation, often in groups, of random ideas, focussed on the problem.

Application by Water Service Providers

The project development processes utilised by a WSP will influence the approaches that might be adopted in the generation of options. For example, in some WSPs the selection of options for wastewater treatment processes is prescriptive, utilising a selection matrix based on simple attributes such as the location of the plant and the population served. This desire for standardisation imposes a pragmatic design approach and clearly inhibits the scope for option generation, and hence the scope for identifying potential alternatives that may be more sustainable in particular circumstances. However, in most organisations, project development processes include value management workshops that are held at some stage of the development process. These workshops will generate alternatives, at least for the components of the system under development. In other organisations creative approaches may be more freely applied, potentially leading to a larger number of alternatives.

In general it is anticipated that the application of creative approaches resulting in the widespread generation of options is more useful where higher level, strategic decisions are being considered, whilst mechanistic approaches are more likely to be useful for decisions relating to detailed design or decisions at an operational level.

The final stages of conceptualisation (option generation) must involve the reduction, by a process of elimination of the options that are considered unsuitable. The options must be checked for compliance to the statement of the problem as an integral part of the conceptual design stage. During this stage, many of the proposals will be rejected as non-compliant and others may be rejected due to a negative performance in relation to technical and financial criteria. The options at this stage will exist in the form of sketches, notes and outline drawings and, although it will be possible to apply quantitative approaches to these decisions, an element of intuitive judgement will still be required and 'weak-decision- making' approaches are likely to predominate. These often occur where decisions to include or abandon options are based on preferences and not detailed information. Some of the decisions at this stage will include constraints that are not readily amenable to inclusion in formalised processes such as SWARD. An example of this may be inappropriate institutional structures for effective accounting for sustainability. Ultimately a reduced number of options that appear to be of sufficient merit at this stage will be taken forward to the subsequent phases of decision-making, supported by approaches such as given here in the SWARD procedure **(Box 3.2 and 3.3)**.

Box 3.2

CASE STUDY

Management of domestic sanitary waste disposed of via WC

PHASE 2 - GENERATE OPTIONS

Six options for dealing with SW were generated for this case study, with an attempt made to encompass the widest possible range of approaches to achieving the objective. The six options address SW discharge at different points in the system, allowing a combination of the alternatives to be used if necessary. The six options approach the problem in three generic ways A: end of pipe solution, B: habit change and C: spill reduction.

The options also define the set of sustainability criteria that should be selected specifically for this case study.

The six options generated for the sanitary waste case study

Option	Measures	Primary objective
1 (A)	Install 6 mm screens on storm overflow at WwTW	To screen storm overflows to constrain solids larger than 6 mm or equivalent to comply with minimum aesthetic pollution requirements for discharges
2 (B)	'Think Before You Flush' (TBYF) campaigns	To encourage a change in domestic disposal habits to dispose of sanitary solids in bin rather than WC.
3 (C)	Install flow storage	To reduce frequency and volume of overflow spills thus reducing the number of SW items discharged to the environment.
4 (C)	Retrofit stormwater source control	To reduce stormwater entry to the sewer system, thus reducing the frequency and volume of overflow spills.
5 (C)	Sewer rehabilitation	To limit infiltration to the sewer system, thus reducing overflow spill frequency and volume.
6 (B)	Retrofit outlet chokes on existing WCs and introduce these to new developments	To force a change in disposal habits from disposal via WC to disposal via bin, due to the increased possibility of WC blockage in the home

Box 3.3

CASE STUDY

Management of domestic sanitary waste disposed of via WC

REVIEW PHASE – DECISION MAPPING

Decision mapping may be conducted to determine the ways in which decisions are made by the WSP, and how sustainability is considered within these processes. This may involve meetings with key personnel, and conducting 'decision-mapping' exercises to follow the steps and criteria used in making particular decisions. Although there are likely to be differences in detail between the ways in which organisations include sustainability considerations in their decision-making, the approaches used are likely to be sufficiently generic that sustainability criteria and a set of DSPs can be produced that would be generally agreeable across a range of stakeholders. Usually a relatively narrow set of economic and technical criteria, such as flexibility of service provision, are the main criteria currently used in decision-making.

In this case study, decision mapping provided confirmation of the validity of the options selected (Box 3), the criteria (Box 5) and the weightings for the relative importance of the criteria (Box 6).

3.3 Phase 3 – Select criteria and indicators

Phase 3 of the decision support process deals with the selection of criteria by which to evaluate the options, generated in Phase 2, as to how their attributes contribute to achieving the objective of sustainability. Criteria, indicators and their relationship to decision objectives are introduced in **Section 1**. This section deals with sustainability criteria; these being defined as the set of factors to be used as part of the overall decision support processes in the assessment as to which of a range of options is likely to offer the greatest contribution to achieving sustainability objectives.

For practical decision-making, criteria that embody and specify the principles of sustainability for the particular decision context must be selected. To aid this process, a generic set of sustainability criteria has been formulated for use in the SWARD DSPs, see **Section 1**. These generic criteria were developed through consultation with water industry partners, research by the SWARD team and review of the academic and technical literature.

This generic set of criteria is based on a hierarchy, consisting of a small number of high-level primary criteria, which cascade down to a larger number of more specific secondary criteria. These primary and secondary criteria are classified under the four primary criteria categories:

- economic
- environmental
- social
- technical.

These categories aim to encapsulate the chosen economic, environmental and social principles of sustainability, together with technical criteria, which relate primarily to the ability of the water/wastewater system to sustain and enhance the performance of the functions for which it is designed and operated.

The primary criteria within each category are given in **Table 3.1**. Under each of these primary criteria, a larger number of secondary criteria are specified. These criteria are informed by indicators (quantitative or qualitative) and hence may be used to assess the future performance of the system under the range of options. The set of secondary criteria and example indicators are described in detail in **Section 1**.

In Phase 3 of the SWARD DSP, a set of specific criteria should be selected which are appropriate to the decision processes under consideration. The generic primary and secondary criteria listed here and in **Section 1** aim to provide a set from which these specific criteria can be selected. This selection process is needed in order to enable evaluation of the relative sustainability of the proposed options, given that some of the generic criteria will be irrelevant or impractical to apply in a particular situation. For any decision process, the set of selected criteria must be broadly applicable to all of the options if comparative evaluation is to be achieved **(Box 3.4 and 3.5)**.

Table 3.1: Set of primary criteria

Category	Primary criteria
Economic	- Life cycle costs - Willingness to pay - Affordability - Financial risk exposure
Environmental	- Resource utilisation - Service provision - Environmental impact
Social	- Impact on risks to human health - Acceptability to stakeholders - Participation and responsibility - Public awareness and understanding - Social inclusion
Technical	- Performance of the system - Reliability - Durability - Flexibility and adaptability

The following aspects of criteria should be borne in mind when selecting specific criteria from the generic list:

Comprehensiveness

The criteria selected should collectively cover the four primary categories of economic, environmental, social and technical, in order to ensure that proper account is being taken of progress towards sustainability objectives. The criteria chosen need to have the ability (via the indicators) to demonstrate movement towards lesser or greater sustainability, according to these objectives.

Applicability

The criteria chosen should be applicable across the range of all the options under consideration. This is needed to ensure that it is possible to compare the sustainability performance of the options.

Tractability

Sufficient reliable numerical or qualitative data or information (indicators) should be available or obtainable to facilitate the assessment of spatial and temporal trends in each of the criteria. Some criteria are not tractable, but may still be considered to be important in terms of assessing relative sustainability. In such cases a WSP will need to explore appropriate methods of gathering data or generating information (i.e. develop indicators) in order for the criteria to be utilised in the future.

Transparency

The criteria should be chosen in a transparent way, so as to help all of the stakeholders to identify which criteria are being considered, to understand the criteria used and to propose any other criteria for consideration.

Practicability

The set of criteria chosen must form a practicable set for the purposes of the decision to be assessed, the tools to be used, and the time and resources available for analysis and assessment. Nonetheless, comprehensiveness or any of the other aspects above should not be sacrificed because of problems of practicability.

The choice of sustainability criteria is crucial, as this will be the primary influence on the outcome of the decision being made, along with the method of comparison or aggregation chosen. The above factors provide initial guidance in the choice of criteria, but how they are to be applied in practice will be built up with experience and review of sustainability assessment across a range of different decisions. **Figure 3.1** shows a suggested procedure for selecting criteria and associated indicators for SWARD decision support processes. There is likely to be a significant element of learning by doing, as experience is built up with the application of the decision support processes given here.

How the criteria are to be used in the subsequent phases of the decision support processes is discussed in the Sections below. Phase 4 deals with the assembly of data for assessing the performance of options according to the criteria, and Phase 5 deals with the analysis of options. This can either be done using a formal multi-criteria analysis or by the comparison of indicators of the various criteria. **Section 1** and **Appendix A** list a large number of example indicators, which were developed along with the criteria in collaboration with water industry and other stakeholders, and from a range of sources including the Water UK sustainability indicators (Water UK, 2000).

It is expected that the generic criteria will be further reviewed, refined and developed by users of the Guidebook as they gain experience. However, we suggest that replacement or non-application of any of the criteria given in the generic list should not be undertaken lightly. This is because:

- The generic list has been produced using research of the SWARD team encompassing the best existing published sources, together with feedback from a wide range of users and stakeholders.
- The generic list includes criteria covering all of the known essential aspects required to evaluate the relative sustainability of options.
- There is very limited experience of applying the concepts given in the SWARD guidebook, and similar decision support processes, hence more experience needs to be gathered before major amendments to the criteria should be made by individual users.

The procedures for the selection of criteria for the evaluation of options should involve the participation of all of the stakeholders affected by the proposed asset management decision. One type of stakeholder involvement process, using a matrix assessment system, is described in the sanitary waste case study in **Appendix D2**.

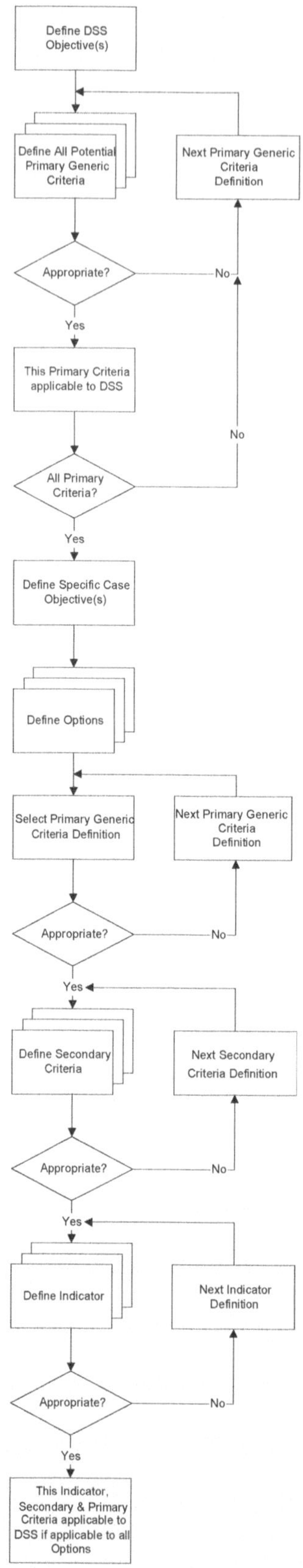

Figure 3.1: Procedure for selecting criteria and associated indicators for SWARD decision support processes

Box 3.4

CASE STUDY

Management of domestic sanitary waste disposed of via WC

PHASE 3 - SELECT PERTINENT CRITERIA

The case study criteria have been selected from (or informed by) the SWARD generic criteria to be relevant to the decision objective and to be applicable to the six options under consideration. In this case study the criteria reflect changes in the system in relation to SW allowing all of the options to be compared.

Primary criteria	Secondary criteria	Indicators
Economic Criteria		
Life Cycle Costs	E1: Capital cost: Investment	£ per catchment
	E2: Operation Costs	£ per catchment per annum
	E3: Maintenance Costs	£ per catchment per annum
Financial Risk Exposure	E4: Financial risk exposure	Qualitative expression of estimated risk
Environmental Criteria		
Resource utilisation	EV1: Energy Use	Total energy (MJ) for 20 year life cycle
Environmental impact	EV2: Impact on air	Total CO_2 emissions (kg) for 20 yr cycle Total NOx emissions (kg) for 20 yr cycle Total SO_4 emissions (kg) for 20 yr cycle
	EV3: CSO/SO discharges to the environment	Volume per annum (cu.m)
Social Criteria		
Acceptability to stakeholders	S1: Acceptability to stakeholders (customers)	Percentage acceptability (Qualitative)
	S2: Perceived impact on the environment	Percentage perceiving negative environmental impact (Qualitative)
Participation and responsibility	S3: Participation and responsibility	Level of participation required for option, High –low (Qualitative)
Technical Criteria		
Performance	T1: Sanitary Waste Escape	Quantity discharged to receiving watercourse per annum (kg)
	T2: Sanitary Waste Transport In Sewer	Amount of SW detained in sewer system per annum (High – low)
Reliability	T3: Risk of Failure to provide service	High – low (Qualitative)
Flexibility and adaptability	T4: Flexibility and adaptability	Ability to accommodate future needs (Qualitative)

Box 3.5

CASE STUDY

Management of domestic sanitary waste disposed of via WC

CRITERIA WEIGHTING

A workshop was held involving eight members of staff from the water authority with the aim of weighting the importance of the selected criteria in the decision making process. The personnel involved were from a wide range of departments. A number of activities were completed in the workshop, involving individual ranking, placing the criteria into a category, and ranking the four categories of sustainability criteria. A group discussion took place as part of the workshop in which the group rankings and categorisation of criteria were agreed. The inter-category importance determined by the workshop participants is shown in the figure below.

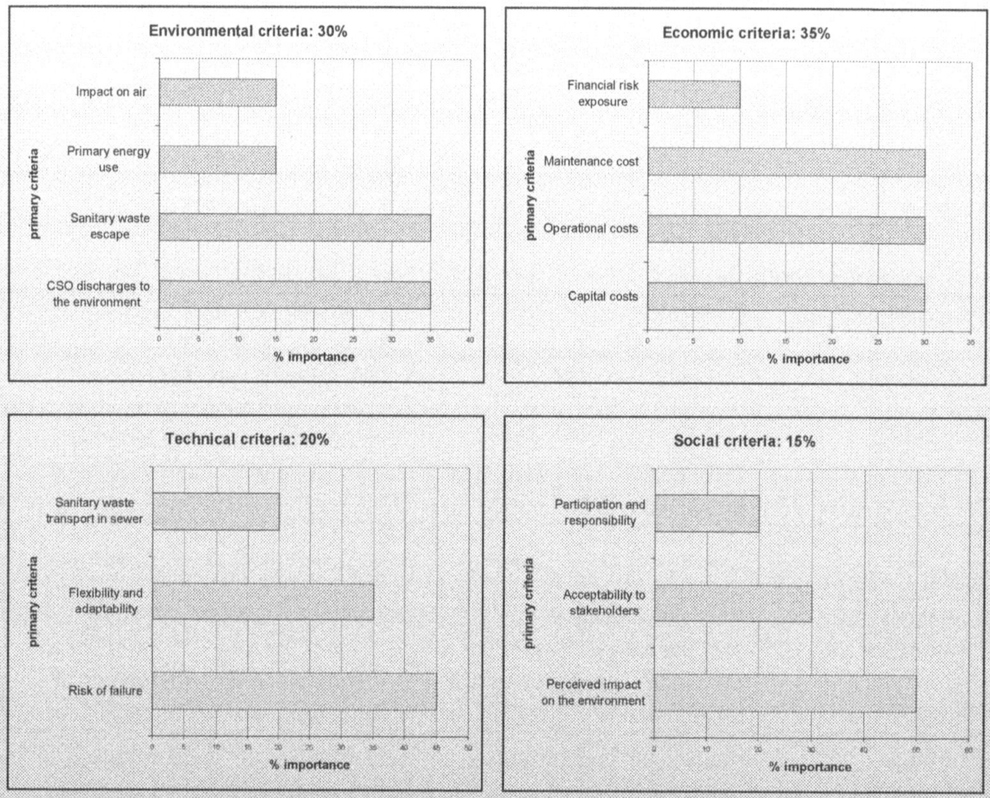

Weighting distributions of the criteria

The weighting distribution of the criteria was based on this collective prioritisation of the SWARD primary/generic criteria. The weighting of each criterion was determined by multiplying the prioritisation factor of each of the four main classification groups to which each criterion belongs to the inter-category importance of the same criterion. For example, it can be seen above that the collective prioritisation of the four groups, economic, social, environmental, and technical, were determined to have 35%, 15%, 30%, and 20% importance respectively. Also, within the technical criteria, the inter-category importance of the 'Risk of Failure' (RoF) criteria was determined to be 45%. Thus the overall weighting of this criteria was determined to be $0.2 \times 0.45 = 0.09$. This was therefore given a weighting factor of 9. These figures were later fed into the three decision-aiding analyses labelled as 'Normal' weights.

3.4 Phase 4 – Collect data and generate information

Two broad types of approach for selecting a preferred option are illustrated in this Guidebook (see **Section 2.6** and **Figure 3.2**). These are:

- comparisons using the pre-defined indicators and criteria; and/or
- formalised Multi-Criteria Analysis

In the first case, options are informally compared according to their likely effects on a range of sustainability indicators. These indicators are those derived from social, economic, environmental and technical sustainability criteria, as described in the preceding sections **(Box 3.6, Box 3.7 and Box 3.8)**.

In the second case, a more formal process is used to analyse options according to a number of sustainability criteria. Both types of decision support processes depend on:

- the collection of indicator data;
- the generation of indicator information.

Information is data that can be readily understood and therefore used to help make decisions. Indicator data, in their raw form, may be suitable for direct use by stakeholders at the decision making stage. Alternatively, they may need to be processed into a more readily understood form (information generation). This is illustrated in **Figure 3.2**.

There are a variety of ways to collect data and/or generate information for relative sustainability assessment. These are briefly discussed below (a more detailed description is given in **Appendix B**).

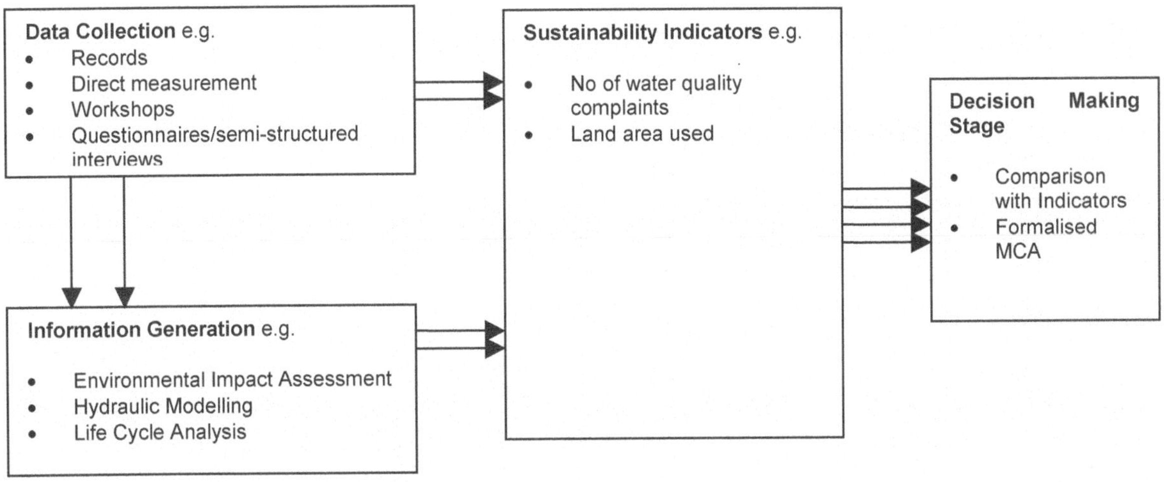

Figure 3.2: Data collection and information generation

Records or direct measurements
The data required for sustainability indicators, or at least some of it, may already have been recorded. It is quite likely that a WSP would keep records of leakage, compliance with consent conditions, production costs or recreational use of watercourses. Such records would be useful in assessing existing systems in terms of environmental, technical, economic and social sustainability criteria respectively.

Some indicator data, not already being recorded, may be collected by direct measurement. Data for the environmental sustainability indicator "Sludge Quality", for example, can be collected by measuring the amount of heavy metals contained in the sludge.

Collecting data from stakeholders
Data can be collected from internal stakeholders (e.g. WSP employees) or external stakeholders (e.g. customers, environmental and community groups).

Data from internal stakeholders will probably be used by the WSP to help understand the effects of changes to waste/wastewater systems. How will changes influence reliability, pumping costs, yield, water quality, pressure etc? Data from external stakeholders is more likely to relate to the acceptability of the effects of changes to water/wastewater systems. How would people feel about a new reservoir, treatment works or pumping station being built near their home? What would the reaction be to domestic water meters, low flush toilets or grey water systems? Will changes be perceived as having a negative effect on health, the local environment, property values, etc?

Ways of collecting data from external stakeholders include:

- Survey research, where:
  - data are collected from a large number of people
  - questionnaires are used, and
  - the data are usually quantitative.
- Focus groups, where:
  - data are collected from a smaller number of people
  - group discussion is used, and
  - data are almost always qualitative.
- Individual 'depth' interviews, where:
  - data are collected from one person at a time
  - the interview is relatively unstructured, and
  - data are qualitative.

Example reference sources suggested for the above data collection methods are those by Strang (2001), which is general, Mangione (1995) for Survey Research, Stewart & Shamdasani (1990) for Focus Groups, and Walker (1985) for individual in-depth interviews.

Box 3.6

CASE STUDY

Management of domestic sanitary waste disposed of via WC

PHASE 4 – COLLECT DATA AND GENERATE INFORMATION

Social criteria

e.g. Criteria S1: Acceptability to stakeholders & S2: Perceived impact on environment

The data for these two 'social' criteria were gathered via a door-to-door survey conducted in the town. Respondents were introduced to the problem of SW escape, and their perceptions and attitude to flushing discussed. The logistics of the six options were explained and illustrated using glossy flash cards. The three alternative approaches to dealing with the problem - end of pipe, habit change and flow reduction – were also outlined and any questions answered. Respondents were asked to rate the six options for tackling SW escape using a range of 1-5. For perceived impact on the environment 1 = *very negative impact* and 5 = *very positive impact*. For acceptability 1 = *not at all acceptable* and 5 = *completely acceptable* to the respondent.

The mean and standard deviation for each criterion were determined and are presented below. The maximum and minimum scores that were possible were also calculated and the combined respondents' scores turned into percentages.

Mean, standard deviation and percentage scores for Criterion S1

S1: PERCEIVED IMPACT ON THE ENVIRONMENT	MEAN	SD	%
1. Install 6 mm screens	3.44	1.15	68.8
2. TBYF campaign	4.13	0.89	82.5
3. Install sewer storage	3.38	1.63	67.5
4. Source control (rainwater barrels)	3.31	1.08	66.3
5. Sewer maintenance	3.88	1.20	77.5
6. Retrofit constricting WCs	3.75	1.53	75.0

Mean, standard deviation and percentage scores for Criterion S2

S2: ACCEPTABILITY TO STAKEHOLDERS	MEAN	SD	%
1. Install 6 mm screens	3.75	1.18	75.0
2. TBYF campaign	4.44	0.81	88.8
3. Install sewer storage	3.50	1.46	70.0
4. Source control (rainwater barrels)	3.06	1.39	61.3
5. Sewer maintenance	4.56	0.51	91.3
6. Retrofit constricting WCs	3.63	1.86	72.5

3.4.1 Tools for generating data

Phase 4 in the decision support process is the collection of data by which to score or compare each option according to the chosen criteria. In order to evaluate the future performance of the system under different options, data collection on the past performance needs to be combined with modelling and assessment of the system over time. There will be an interaction between the criteria chosen for assessment and the modelling tools used to evaluate future performance of the system. Detailed modelling tools are essential to capture the performance of the system under specified conditions relating to each option under consideration. The tools chosen should be appropriate to model the future values of the criteria under the range of options to be considered.

The water industry is familiar with many of these tools, particularly those such as environmental impact assessment and cost-benefit analysis. However, traditional approaches such as these have major deficiencies when sustainability is being assessed. In order to properly evaluate relative sustainability criteria for all of the primary areas: environment, society, economics, and (for infrastructure) technology must be assessed. Without consideration of a wide enough range of criteria covering all of these areas, key elements of sustainability will be ignored. For example, traditional environmental impact assessment (EIA) often ignores many important socio-economic criteria, whereas cost-benefit analysis traditionally deals only with 'local' costs and benefits and not economic or whole-life-costs (Penning-Rowsell *et al.*, 2003). There are signs, however, that even these techniques are beginning to broaden their perspectives, with EIA professionals starting to take socio-economics more seriously (Chadwick, 2002). A detailed analysis of the range of tools that are available for the generation of relevant data is given in **Appendix B**, brief summaries of these tools are given below.

Systems analysis tools

As there are a hierarchy of different types of model available often a combination of different levels of modelling will be required. These range from detailed process modelling tools, such as InfoWorks™, to higher-level systems modelling approaches, such as the Reference Sustainability System (see below). The interaction between different levels of modelling is a key feature of the assessment process. Sustainability assessment requires a systems based approach and whatever model system is used it should contribute to this perspective. A good example of where a systems approach is being taken to sustainable water management is as part of the Swedish Urban Water programme (Urban Water, 2001). Systems analysis in this programme refers to the interdisciplinary research approach to the comprehensive analysis and assessments of sustainable urban water systems. The Urban Water approach is to develop criteria for sustainable water and wastewater systems, reflecting the multi-disciplinarity needed for comprehensive understanding and analysis. Models and assessment methods are being developed and tested for each criteria group, representing the fundamentals of the systems analysis used in the programme.

For example, in the sanitary waste case study, an InfoWorks™ model was built to simulate the dry and wet weather flow in the combined sewer system under investigation.

However, this type of model is highly data intensive and often requires long computational run times, making it impractical to do more than a limited number of model runs. Thus, it may be necessary to supplement detailed modelling tools with higher-level systems models, which allow more flexible assessment of the interaction between different parts of the system under a wider range of conditions. This type of model is exemplified by the Reference Sustainability System approach.

The Reference Sustainability System (RSS) approach has been developed by members of the SWARD research team at Imperial College (Leach *et al.*, 1997; Foxon *et al.*, 1999,

2000). This was a development of the original Reference Energy System (RES) approach, which is widely used for modelling energy systems (Meier, 1984). The Reference Sustainability System is a methodology for representing the resource and material flows on which the environmental sustainability of a system depends, in order to enable systematic assessment of the potential of technologies and resource management strategies to enhance the sustainability of the system. An RSS systems model aims to represent the key activities over the life cycle of a resource flow, from source, supply, and distribution to end-use and waste disposal. Constraints are imposed on the model to ensure that materials balance is maintained at each stage of the system (ensuring all flows are accounted for), that end-use demands are met, and the technical capacity of systems elements is not breached. A range of options and scenarios for technological or management changes to the system can then be modelled by changing the constraints, and solving the model for the flows which meet the newly-constrained system.

Such a systems model can either be developed as a stand-alone model, requiring the collection of large amounts of data, or as a complement to a more detailed process model. In the latter case, the systems model can take the output of the finite runs of the detailed process model as a starting point and interpolate the state of the system for values between these points. Whilst this introduces another level of approximation into the modelling, it can nevertheless be useful for understanding the system behaviour under a wider range of conditions. For example, the systems model can be used to assess the results of taking combinations of the basic options, in order to estimate the optimum combination according to specified criteria. This estimated optimum combination can then be tested again using the detailed process model. This type of modelling was applied in the sanitary waste case study, as described in **Appendix D2**.

Social Analysis Tools

There are a range of methodologies and tools that can help with the incorporation of social data into the decision-making process. These include data collection techniques such as households surveys, questionnaires, focus groups and workshops as described above, through to the more structured analytical frameworks such as Social Impact Assessment (SIA) and Social Capital Assessment Tool (SOCAT). These frameworks can help to assess the likely impact of policy, or other decisions, on stakeholder groups, and also examine the effect that social networks and behaviour may exert on a particular decision outcome. In recent years computational approaches that use Artificial Intelligence (AI) to study social organizations and institutions have become increasingly common. A number of modelling approaches can be used to depict social networks, with the most common contemporary approach being the use of agent-based modelling (ABM). Some of the various tools are described in more detail in **Appendix B**.

Economic analysis tools

Alternative investment options will almost certainly be subjected to some forms of economic analysis. Thus, the information generated by economic analysis tools will very often have a significant bearing on the final decision. Three important analysis tools are briefly discussed below and described in more detail in **Appendix B**:

***Cost-benefit analysis* (CBA):**

- CBA used to determine if the benefits of an investment outweigh its costs;
- there are two types of CBA: economic and social;
- economic CBA involves real 'private' benefits and costs that affect the investor;
- social CBA replaces private benefits and costs with social costs and benefits, i.e. beyond those of direct interest to the investor (e.g. social and environmental externalities);
- in economic CBA the costs and benefits of alternative options are identified and quantified in monetary units (monetised);

- in social CBA, externalities are also usually monetised and included in the analysis, and 'distorted' market prices (e.g. distorted by subsidies or market imperfections) may be adjusted to 'shadow' prices;
- future costs and benefits are normally discounted to arrive at a present value (PV);
- subtracting discounted costs from benefits gives the net present value (NPV);
- NPV is a commonly used selection criterion, the higher the NPV the better;
- An alternative criterion is the Internal Rate of Return (IRR), i.e. that rate of discount which yields a zero value of NPV

From a sustainability viewpoint, social CBA is preferable to economic CBA because its scope is wider and the time horizon may be longer when taking account of social, environmental and economic factors and does not just consider the costs and benefits from the viewpoint of the individual actor concerned. It encompasses three different aspects of cost benefit analysis; the effects on all members of society, the distributional effects and, market distortions and imperfections. However, the monetisation of externalities is not without its problems.

Cost-effectiveness analysis **(CEA)**

- variant of CBA;
- used as a comparison where outputs of alternative projects are equivalent;
- as project benefits are common, they need not be monetized, but can be expressed as units of output;
- projects judged according to their discounted cost (cost-effectiveness);
- the lower the cost the better.

CEA has the disadvantage of being limited to situations where alternative projects have different costs (monetary inputs) and equivalent (non-monetary) benefits. It is probably more common for alternative projects to have different effects – beneficial or otherwise – even though they fulfil the same basic function.

Whole life costing **(WLC)**

- WLC used to determine whether a potentially higher initial capital cost of an investment is justified by lower costs elsewhere, or if the life expectancy of a chosen component is suitable for a chosen period of study;
- involves the systematic consideration of all relevant costs and revenues associated with the acquisition and ownership of an asset;
- includes both initial capital or procurement costs, maintenance and operation costs, opportunity costs, future costs and externality costs;
- all costs are identified and quantified in monetary units;
- options with lower costs over the period will be preferred;
- uses NPV to enable comparisons to be made between options.

WLC is being increasingly advocated by the UK government as a way of optimising investment decision-making and to ensure the adoption of cost effective solutions. The technique is finding wide application in fields such as transportation planning, built environments, higher education facilities and water service infrastructure. WLC requires the consequence to be defined in terms of the absolute value (in monetary units) and requires robust performance and accounting modules to be in place in order to assign consequences to actions. Use of WLC by the water industry has the advantages that: the process is fully auditable to external reporters; meets the requirements of Ofwat's serviceability aspirations (Ofwat; Byatt, 2000; Heywood *et al.*, 2003) in providing a holistic and robust rationale for asset management; considers both costs and performance over extended periods; and incorporates issues related to resource use and impacts on the environment and society.

Even where it is agreed that externalities should be monetised, it is often very difficult to do so. It may be more realistic to admit that some externalities defy monetary translation, which has serious implications for CBA, CEA and WLC (Skipworth *et al.*, 2002). Ding (1999) suggests that these unpriced, but still important, impacts be included descriptively in a CBA report. Decision makers can then determine the weight to be given to these issues in their policy determination. Similarly, Merret (1997) proposes a dual analysis approach where CBA is combined with Environmental Impact Assessment.

Box 3.7

CASE STUDY

Management of domestic sanitary waste disposed of via WC

PHASE 4 – COLLECT DATA AND GENERATE INFORMATION

Economic criteria

Semi-structured interviews were conducted with a range of different levels of personnel from the water authority to gather data for the economic criteria. Information was required on processes, maintenance protocol and plant design, with identification of the revenue costs directly attributable to SW in the waterborne system for the town. The presence of SW material in the waterborne system adversely affects the operation of certain components within the WwTW. Much of the indicator data consisted of measurements from meters installed at the treatment works.

E.g. criterion E1: Capital Costs

For each investment, capital costs were collected directly from the suppliers of the equipment and materials. The cost of installation were estimated and included in the final cost. Costs of contractors and externalities such as excavation were also included. The capital costs for each option were as detailed below.

Option		Details
1	Install 6 mm screens on overflows at WwTW	The cost of a drum screen, including installation, was £68,000. The screen model was selected for the appropriate flow volume from the screen suppliers of the WwTW.
2	Think Before You Flush' (TBYF) campaigns	The campaign for the catchment population cost £10,000. This includes materials, transport and personnel costs.
3	Install flow storage	Sewer storage cost £70,000 for excavation, materials and construction.
4	Retrofit stormwater source control	157 rainwater barrels at £25 each are required for the target reduction. The total cost for the catchment for purchase and delivery is £3295.
5	Sewer rehabilitation	This requires the replacement of 78 m of sewer at varying diameter and depth. The average cost is £500 per meter = £39 000 for excavation, materials and construction.
6	Retrofit outlet chokes on existing WCs and introduce these to new developments	This involves the retrofitting of 522 WCs in the catchment. Materials and installation costs are £43 per WC, with an additional £10,000 for the complementary educational campaign required.

Environmental analysis tools

Numerous analytical tools exist that help provide information about the environmental consequences of a decision.

The European Network on Chain Analysis for Environmental Decision Support (CHAINET) project has described, compared and assessed ten of these tools (Wrisberg *et al.*, 2002):

1. Life Cycle Assessment (LCA)
2. Material Input per Unit of Service (MIPS)
3. Environmental Risk Assessment (ERA)
4. Material Flow Accounting (MFA) a.k.a Mass Balance Analysis
5. Cumulative Energy Requirements Analysis (CERA)
6. Environmental Input-Output Analysis (env.IOA)
7. Analytical Tools for Eco-design (Eco-design Tools)
8. Life Cycle Costing (LCC)
9. Total Cost Accounting (TCA)
10. Cost-Benefit Analysis (CBA)

Their potential application within the SWARD DSPs is outlined in **Table 3.2**. The main features of the tools given in the table are described further in **Appendix B**. Further information can be obtained from the CHAINET book (Wrisberg *et al.*, 2002).

Table 3.2 Analytical tools that can help provide information about the environmental consequences of a decision (Adapted from CHAINET, Wrisberg et al.*, 2002)*

Tool	Focus	Data requirements	Strengths	Weaknesses
LCA	Environmental impacts over life cycle of product or service	Very high	Relates to function Comprehensive Cradle-to-grave Quantitative or qualitative	Collection of data Where to draw boundaries for analysis
MIPS	Relates material input over life cycle to final service delivered to user	High	Quantify material intensity of products/services Identify key material inputs	Collection of data Definition of end-use service
ERA	Hazard identification Effects assessment Exposure assessment	Very high	Comprehensive evaluation of the probability that damage or adverse effects will occur	Determination of effects of & exposure to a range of chemicals
MFA	Accounts in physical units for flow of materials through a specific region	High	Provides info on contribution to sustainability of bulk or specific material flows	Needs to be supplemented by assessment of environmental impacts of flows
CERA	Quantify primary energy requirement for products & services in a life-cycle perspective	High	Quantify energy intensity of products/services Identify options for energy savings	Needs to be supplemented by assessment of environmental impact of energy generation/use
Env. IOA	Display all flows of goods & services, & the inter-connections between them	Very high	Allows consistent treatment of environmental & economic factors	Highly data intensive Does not focus only on key components. Assumes linear production relationships
Eco-design Tools	Matrices relating different life cycle stages to extraction, intervention, or emissions Checklists for key questions or points of attention	Moderate	Relatively quick & easy means of identifying key environmental impacts	Lacking comprehensive assessment of environmental impacts
LCC	Identify all internal & external costs over the life-cycle of a product or service	Very high	Puts a monetary value on emissions & resource use quantified in life cycle inventories, & environmental & health effects in impact assessments	Practical & conceptual difficulties in assigning monetary values to environmental impacts
TCA	Long-term, comprehensive analysis of full range of internal costs & savings from pollution prevention & environmental projects	Very high	Comprehensive costs & savings inventory More precise cost allocation Long time horizons Use of profitability indicators	Difficult to identify & assess all long-term costs & savings
CBA	Economic tool for supporting decisions on larger investments from a social, rather than a firm's point of view	Very high	Can take into account costs of externalities, i.e. effects which are not expressed in market prices	Practical & conceptual difficulties in externality costings

Box 3.8

CASE STUDY

Management of domestic sanitary waste disposed of via WC

PHASE 4 – COLLECT DATA AND GENERATE INFORMATION

***Environmental Criteria* e.g. Criteria EV1: Energy Use & EV2: Impact on air**

The use of Life Cycle Assessment (LCA) is a key component of the case study, and is used to generate data for two of the environmental criteria. The life cycle of each option is assessed individually identifying: materials used; energy consumed; and releases to the environment, encompassing transportation, recycling and re-use. Assumptions are made in relation to the design life of a particular product and the likely replacement parts requirements. The main sewer network and treatment works (other than screens) have not been considered in the analysis, as the infrastructure is the same for each option. Only new infrastructure associated with the control of sanitary waste is investigated. The options are compared in relation to a number of indicators from the LCA (SimaPro™ - PRé consultants B.V.,1997) package. This allows an assessment of the relative impact (part of the relative sustainability) of the six alternatives. A number of sub-options were assessed for options 4 and 5. Sub-options 4a, 4b and 4c relate to 100%, 35% and 28% of houses in the town being fitted with rainwater barrels. Sub-options 5a through to 5f relate to the length of sewer rehabilitated in the town - 100%, 72%, 42%, 30%, 18% and 11% respectively. The data have been collected from various sources. The manufacturer of the products used for each option were identified and contacted via phone, fax and email, as well as site visits. The processes involved in the manufacture were identified and data on the manufacturing process obtained. Where data were not available for a specific process the database was supplemented by data from the SimaPro™ database. Data have been collected based on the following material components required for the various options:

Material components required for the six options

Option	Components
1	6 mm drum screen at the storm outfall with 310 l/sec screening capacity
2	Leaflets and questionnaires
3	Concrete tank (60m * 4m * 5m) to store 1100 m^3 of water
4	Plastic barrels, the number of barrels used with sub-options (a, b, and c) are 626, 219 and 175 respectively.
5	Replacing 300 mm, 450 mm and 600 mm pipes with different lengths for each sub-option. The pipes lengths for sub options (a, b, c, d, e and f) are: 144m, 108m, 87m, 54m, 42m and 30m respectively.
6	Replacement of the toilet outlet connection with a smaller plastic pipe of 75 mm diameter and the cistern-flushing valve to reduce the flush volume to 3 l/flush.

In order to compare the projected total emissions from the options a functional unit has been selected - the gross solids reduction during the option life cycle (taken as 20 years for the purpose of this case study). The total emissions of CO2, SO4 and NOX, along with the total energy use over the life cycle of each option are presented in the table below:

Emissions values per kg gross solids reduction

Option	Gross solids reduction in kg/20 years	Greenhouse gases kg CO_2 equ.	Acidification kg SO_4 equ.	NO_x kg	SO_2 kg	Energy MJ
1	2046.017	8.016	0.207	0.009	0.176	127.076
2	1416.453	0.569	0.003	0.000	0.000	11.790
3	1099.604	136.413	1.719	1.710	0.316	1282.279
4-a	959.016	18.248	0.111	0.009	0.016	535.966
4-b	369.509	19.567	0.120	0.010	0.017	584.560
4-c	303.404	16.018	0.097	0.008	0.014	471.320
5-a	1262.132	4.683	0.022	0.012	0.001	59.265
5-b	1056.150	4.090	0.019	0.010	0.001	51.887
5-c	747.654	4.186	0.020	0.011	0.001	53.099
5-d	595.259	3.444	0.017	0.009	0.001	43.678
5-e	426.709	3.843	0.019	0.010	0.001	48.745
5-f	269.049	4.200	0.021	0.010	0.001	53.522
6	2023.009	2.269	0.030	0.015	0.003	73.158

3.5 Phase 5 - Analyse Options

The fifth phase of the SWARD approach is to analyse the relative sustainability of the range of options being evaluated **(Box 3.9)**. As explained in **Section 2** the very nature of such decision problems is that they are usually complex: the decision options are often multi-attributed; there are usually several stakeholders/decision-makers, each one seeking to satisfy several decision criteria according to their own (or their perception of their organisation's) preference structure. In order to assist in the evaluation of the range of economic, social, environmental and technical criteria, a number of decision-making approaches and tools can be applied. This section of the guidebook describes how a number of the more important decision-making approaches and tools can be applied in the context of assessing the sustainability of water/wastewater systems.

Section 2.5, and its supporting **Appendix C**, focuses mainly on the selection and use of analytical decision-making tools.

Box 3.9

CASE STUDY

Management of domestic sanitary waste disposed of via WC

PHASE 5 – ANALYSE DATA

The models and weightings used

The data for all 16 of the case study criteria were collated and three MCA models - ELECTRE III, PROMETHEE and SMART - used in phase 5 to analyse the full set. The ranking and weighting of the criteria obtained from the workshop were combined with the information on the preferences thresholds, indifference thresholds and veto thresholds for each criterion. The weighting of each criterion was determined by multiplying the prioritisation factor of each of the four main classification groups to which each criterion belongs to the inter-category importance of the same criterion. The collective prioritisation of the four groups - economic, social, environmental, and technical - were determined to have 35%, 15%, 30%, and 20% importance respectively. For example, within the technical criteria, the inter-category importance of the 'Risk of Failure' (RoF) criteria was determined to be 45%. Thus the overall weighting of this criterion was determined to be 0.2 × 0.45 = 0.09, and was therefore given a weighting factor of 9. This was the 'normal' weighting system.

The ranking of the options suggested by each of the methods of analysis is presented below:

Option	PROMETHEE	ELECTRE	SMART
1. Install 6 mm screens	5th	6th	4th
2. TBYF campaign	1st	2nd	1st
3. Install sewer storage	6th	5th	6th
4. Source control (rainwater barrels)	3rd	=3rd	2nd
5. Sewer maintenance	2nd	1st	3rd
6. Retrofit constricting WCs	4th	=3rd	5th

Overall consideration of the results would suggest that the TBYF option generally dominates the top rankings, followed by sewer rehabilitation and then the source control option. The other three options are quite consistently placed in the final three positions. However, whilst there would appear to be a reasonable degree of consistency between the outranking methods (PROMETHEE and ELECTRE), there is less consistency between these methods and the SMART analysis. This is likely to be due to the influence of trade-offs in the smart analysis between positive and less positive scores for individual criteria within the analysis.

3.6 Phase 6 – Selection of preferred option

The analytical tools described in **Section 2.5** provide information to guide the decision-maker about the relative sustainability of the options, but a further stage is required in the analysis before final option selection. Analytical approaches do not remove the need for human judgement on issues such as risk and uncertainty and on the influence of intangible issues (those that defy monetary assessment) on the final decision.

The process of selection of the preferred option is shown in **Figure 3.3**

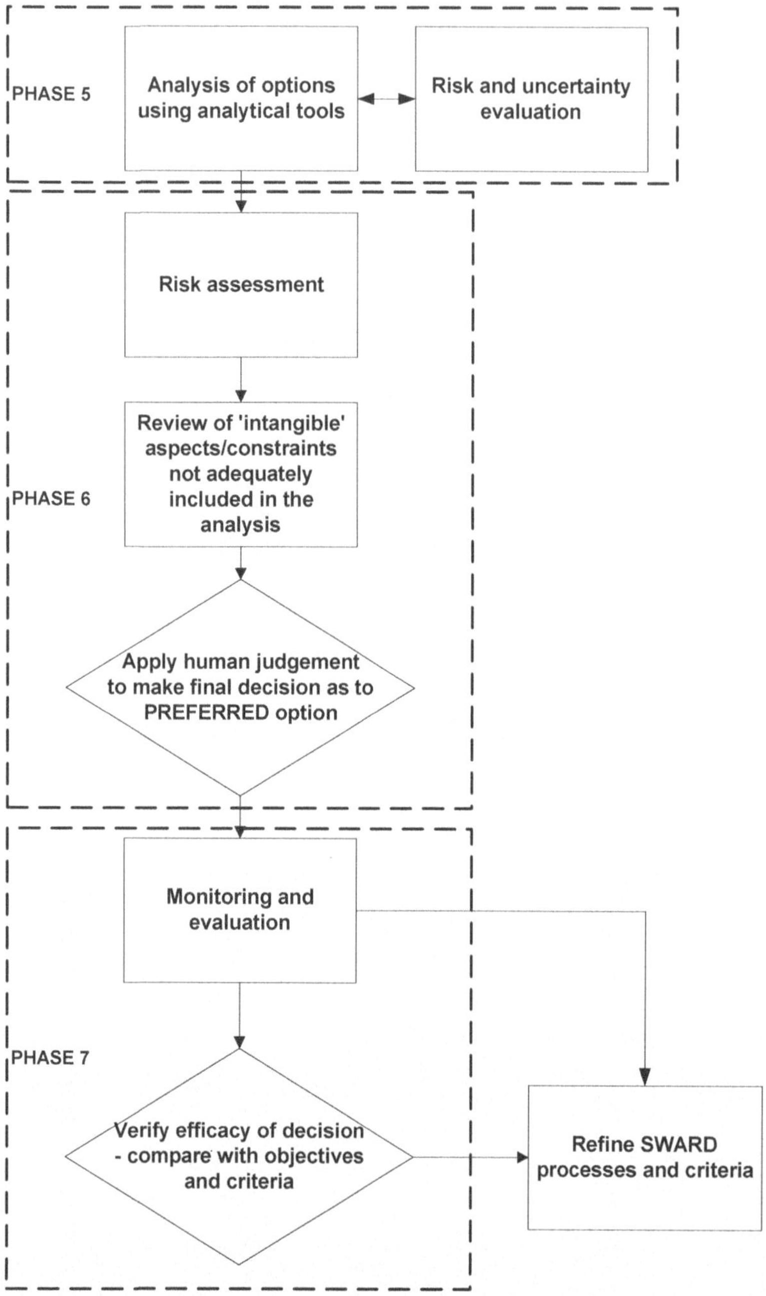

Figure 3.3: Selection of the preferred option

3.6.1 Risk and Uncertainty

The application of the techniques above produces data that establishes whether one option is more or less sustainable than another. However, before completing the analysis it is essential that the decision-maker should reflect on the relative ratings of the options before making a final decision. Consideration must be given to the possibility that a wrong decision might be taken due to inaccurate data, much of which will be subjective, and to the inherent simplifications within the techniques. An appropriate risk management process is as follows with a full evaluation of the role of risk analysis within the overall SWARD framework given in **Appendix C.**

Risk Identification

In this context, the risks arise due to imprecision in establishing the scores and weights for criteria. It is therefore necessary during the data collection stage to establish the degree of certainty of the values estimated for these by recording the range of values or opinions obtained.

Risk Analysis

Standard risk analysis tools such as sensitivity analysis or monte-carlo simulation can be applied to MAUT and SMART analysis.

Risk Response

The risk of making a wrong decision can be minimised by avoiding options where the sustainability is highly uncertain. If this was not possible or not desirable, then it would be essential to devote more resources and more sophisticated methods to the assessment of key indicators to limit the risk of a score being wrong.

Examples on the application of sensitivity analysis and risk analysis to decision analysis tools are given in the sanitary waste case study in **Appendix D2**.

3.6.2 Incorporation of Intangible Aspects

In practice, it will be very difficult (and in some cases impossible) to obtain credible weights and scores for all aspects of a sustainability decision. Increasing inclusion of intangible aspects will decrease reliability of the results of any decision analysis and such aspects should be excluded from the quantitative forms of analysis described in **Section 2.5** and in the sub-sections above. It is more realistic to admit that some aspects are not compatible with quantitative analysis and to list these for consideration in a final qualitative evaluation stage of the decision process.

3.6.3 Making a Decision (Box 3.10 and Box 3.11)

At this point, the decision-makers will consider each option in terms of the results of the decision analysis, the risk assessment of the robustness of this analysis and the qualitative information on intangible aspects. Ultimately they must use human judgement to make the final decision on which option is the more sustainable. This requires the decision-maker to make a choice based on experience, values, perception and intuition. The analytical process should overcome a number of cognitive limitations of the human decision maker, in that the number of pieces of information will have been reduced to manageable numbers and the risks associated with decision (in terms of relative sustainability) will have been identified. Ultimately the decision-makers must use their human knowledge and experience to make a final judgement on which project or combination of projects is likely to be the more sustainable.

Box 3.10

CASE STUDY

Management of domestic sanitary waste disposed of via WC

PHASE 6 - SELECT PREFERRED OPTION

Decision makers may not select the option that has been shown to be the more sustainable. This may be due to a number of reasons; leading to constraints that could not be fully included in the MCDA. Nonetheless due to the openness of the process, where the most sustainable option is not selected, other stakeholders will be aware of this.

Risk Assessment: In the context of the decision making process, the risk of reaching a wrong conclusion arises from variability and inaccuracy in the weights and scales used in the analysis in Phase 5.

MCA Sensitivity Analyses: Sensitivity analyses were performed by changing the value of the weights applied to the criteria. These were conducted so that the overall outcome of the project could be considered in terms of subjective risk and the robustness of the results. The MCA sensitivity analyses were carried out by inverting the (normal) weighting function of each of the criteria, and also by running the three models using an equal weighting of the criteria.

Results of the MCA

Option	PROMETHEE			ELECTRE			SMART		
				Weighting					
	Normal	Inverse	Equal	Normal	Inverse	Equal	Normal	Inverse	Equal
1	5th	5th	5th	6th	=4th	5th	4th	3rd	4th
2	1st	1st	1st	2nd	1st	=1st	1st	1st	1st
3	6th	6th	6th	5th	6th	6th	6th	6th	6th
4	3rd	2nd	3rd	=3rd	3rd	=3rd	2nd	2nd	2nd
5	2nd	3rd	2nd	1st	2nd	=1st	3rd	5th	3rd
6	4th	4th	4th	=3rd	=4th	=3rd	5th	4th	5th

MCA results: The full set of results for the MCA illustrate that in terms of sensitivity, despite the wide variance of the weighting, the general inference of the rankings appears to be quite robust for all three analyses. Options 2, 4 and 5 (TBYF, retrofit source control and sewer rehabilitation respectively) generally occupy the top ranks for all models. Options 1, 6 and 3 (install 6 mm screens, fit constricting WCs and install flow storage) generally occupy the lower three ranks. For the inverted weighting set, TBYF dominates the highest rank across all three analyses. The source control option is quite robust, appearing as second rank for two of the analyses. Finally, for the equal weighted set, the top places are generally constant, with TBYF, source control and sewer rehabilitation in order of general preference. Installing 6mm screens, fitting constricting WCs and installing storage are unwavering in the fourth, fifth, and sixth places that they occupy and can therefore be considered to be less sustainable solutions according to the criteria used. Further risk assessment was undertaken on the SMART analysis using Monte-Carlo simulation with the scores for criteria being replaced by probability distributions. Based on the overall results of the three models, and the sensitivity testing, the final ranking can be considered as that shown in the first two columns below:

Overall ranking of options

Rank	Alternative
1	TBYF
2	SeRe
3	RSSC
4	ROC
5	In6S
6	InSt

Decision makers ranking

Rank	Alternative
1	In6S
2	InSt
3	SeRe
4	RSSC
5	TBYF
6	ROC

It is the decision-maker's responsibility to consider if in selecting the TBYF option, the assumptions made about public changes in habit are not too optimistic. The final decision depends very much on the attitude to risk of the decision-maker, i.e. risk-taker, risk-neutral or risk-adverse. Both 'investment' options – sewer rehabilitation and retrofitting source control - are very close in terms of their utility and can complement the TBYF option in the case of an adverse to risk or neutral to risk decision-maker. The MCA ranking is in a stark contrast with the one established on the basis of professional judgement (with no MCA) by the decision-makers in a workshop (see above). This clearly shows that the decision-makers have top ranked the poorest options and emphasises the necessity to undertake MCA, especially for strategic decisions.

Box 3.11

CASE STUDY

Management of domestic sanitary waste disposed of via WC

PHASE 6 - SELECT PREFERRED OPTION

Making the Decision – the application of human knowledge and experience

In order to simulate the final decision making stage, the results of the MCDA and the risk analysis were presented to a group of water industry experts with different professional backgrounds. The group were asked to review the data provided and to apply their judgement and experience to facilitate the inclusion of any intangible aspects not yet considered; and to reach a final decision.

An initial discussion focussed on the identification of any intangible issues not yet included in the analysis. A number of issues arose; some already had been included in the selection of criteria. These were:

- That options selected will be constrained by the need for the water industry commissioner to buy in to any novel or alternative approach to tackling a problem.
- That flexibility in the selection of options will also be constrained by an expected budget spend which may have been set earlier in the decision making process. This may in some cases limit option choices, but it was noted that this could have been included in the ratings of the economic criteria.
- There was some concern about selecting options where high levels on public participation were required. This is present within the decision making team even though this issue has been taken into account by the criteria. It is seen as one of the key overriding issues considered.
- Regulatory impediments are a problem when trying to select the best option. The group were not certain that it be possible to get agreement from regulators to remove a screen by implementing another option.
- In general the group identified a lack of confidence in selecting unproven methods, this issue exists in the decision makers minds as well.

The final selection

Following a discussion it was decided that a combination of unproven methods - which were generally identified by the MCDA analysis as being more sustainable - and a complementary proven method was the most appropriate selection. The public relations value of the TBYF and source control options were noted and deemed to be complementary. When houses in the town were visited in order to deliver rainwater barrels the TBYF educational campaign could also be undertaken.

3.7 Phase 7 - Post decision phase: Monitoring and Feedback

This section deals with the monitoring and evaluation of the decision on implementation of the selected option. This phase may necessitate monitoring over a significant period (possibly decades, depending on the nature of the development).

Monitoring and evaluation has a multi-purpose, firstly, it should confirm that the evaluation undertaken in the decision stage was realistic; this is the decision review stage. Secondly, the process review stage provides a means to refine the generic and specific criteria, the database and indicators (where used) and the SWARD processes **(Box 3.12)**.

3.7.1 Decision Review

The purpose of decision review is to verify that the selected option is that which contributes the most to enhancing sustainability. Unfortunately without implementation of each of the candidate options it may not be possible to ascertain this with much certainty. It should be possible, however, to review the selected criteria, assumptions, weightings and risk assessments used for the selected option. In addition, the collection of data about option performance will allow assessments to be made of the indicators and their temporal changes.

Figure 3.4 provides an overview of the approach that may be used for this. A major problem will be to ensure both robustness of the data collected and also not to require too onerous resources for this activity. Examples of approaches to data collection for (natural) water quality assessments are given by Chapman (1996) and for post-project appraisal at low cost by UKWIR (2001).

3.7.2 Process Review

The SWARD DSP given in this Guidebook is dynamic. It will require continual re-appraisal and revision for a number of reasons, examples of these may be:

At a conceptual/strategic level:

- As the understanding of sustainability is enhanced and how best to include this in future developments.
- As constraints alter with time – legislative and corporate attitudes etc.
- As decision support techniques advance, this may provide new opportunities and approaches.

At a decision level:

- As users develop the way in which their objectives are defined and implemented.
- As criteria become more refined – experience with use of the proposed criteria will demonstrate their robustness and suggest alternatives.
- As stakeholder inclusion and the decision formulation process becomes more effective.

The approach to this is informed by the decision review above (**Figure 3.4**), but will also include a much broader scoping process, to ensure that latest knowledge is incorporated in the SWARD processes.

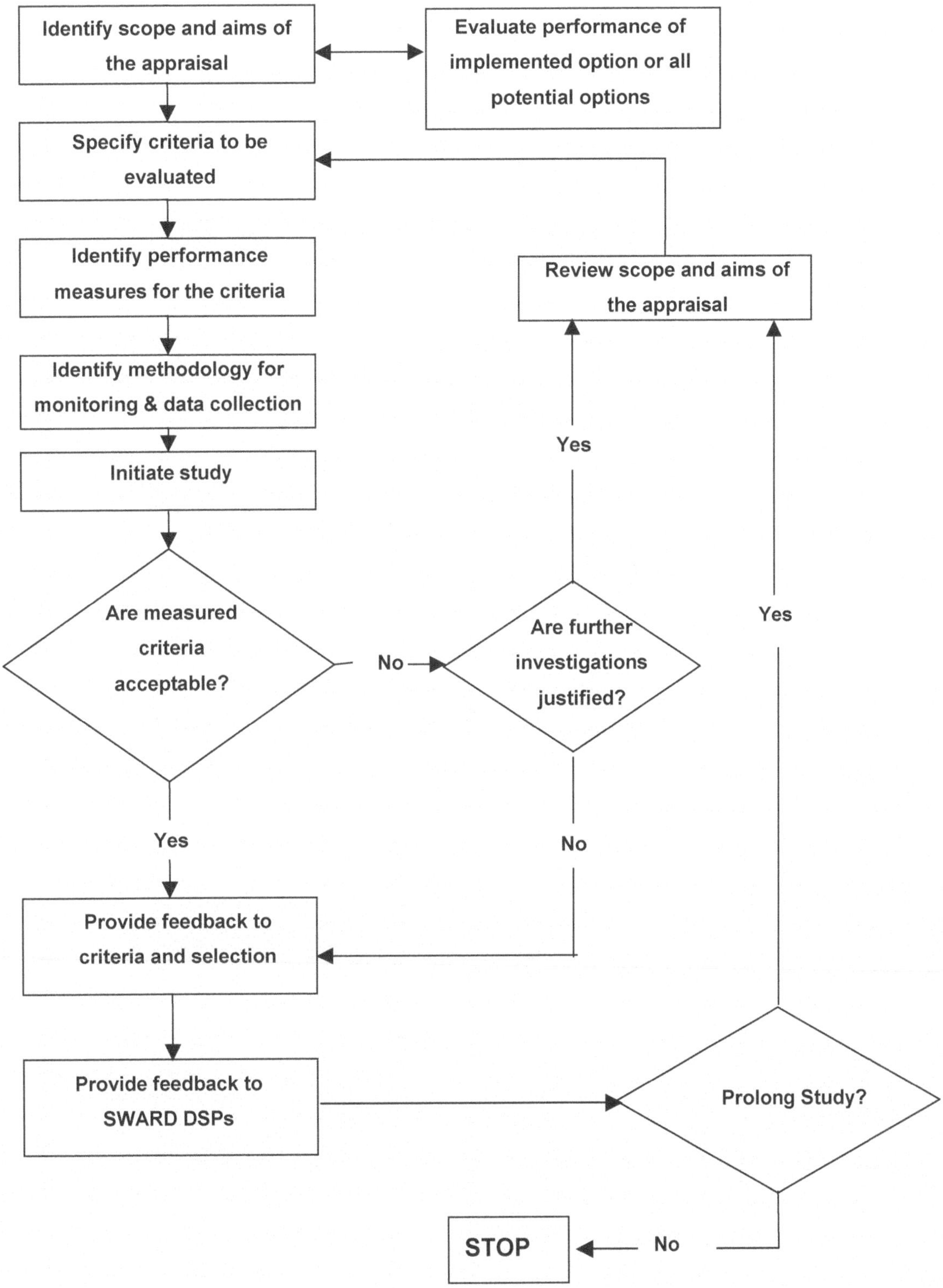

Figure 3.4: Post-implementation performance appraisal

Box 3.12

CASE STUDY

Management of domestic sanitary waste disposed of via WC

PHASE 7 - POST PROJECT MONITORING AND FEEDBACK

Indicators and potential for measurements following implementation of selected option

Criteria	Units	Change to enhance sustainability	How determined in initial analysis	Opportunities for actual measurement post implementation
T1: Sanitary Waste Escape	kg/yr	Reduce	Amounts of solids originating based on social surveys & checked with screen capture. Escape estimated by pro-rata volume splits for storm overflows using sewer flow simulation model.	1. Carry out TBYF surveys to verify inputs of SW (EV3). 2. Monitor flows to verify flow & volume splits at CSOs. 3. Undertake beach & marine surveys of SW.
T2: SW Transport	Qualitative	Reduce	This was assessed in terms of the propensity to be detained in the sewer network by qualitative means.	This could be assessed from the hydraulic model with depth & velocity thresholds*, or by CCTV in-sewer inspection.
T3: Risk of failure	Qualitative	Reduce	Qualitative assessment from expert discussion.	This could be determined with confidence only from long-term data over several years as to the overall performance.
T4: Flexibility & adaptability	Qualitative	Increase	Qualitative assessment from expert discussion.	As above, & only following the imposition of new constraints or introduction of new innovations.
E1: Capital cost	£	Reduce	Actual estimates from suppliers etc.	The real as implemented costs should be available.
E2: Operational cost	£/yr	Reduce	Estimated from suppliers & historical data	Collect data over the first year of implementation
E3: Maintenance cost	£/yr	Reduce	As above	As above
E4: Financial risk exposure	Qualitative	Reduce	Qualitative assessment from expert discussion.	Following implementation this may only be further assessed by comparison with other similar implemented schemes.
S1: Acceptability	%	Increase	Social survey	Follow-up social survey
S2: Perceived impact on the environment	%	Reduce	As above	As above
S3: Participation & responsibility	Qualitative	Increase	As above	TBYF survey or equivalent to assess actual public 'take-up' & involvement
EV1: Energy use	MJ	Reduce	LCA	Collect data over the first year of implementation for direct energy use. Energy in materials and recycle potential not readily measurable.
EV2: a) Impact on air (CO_2)	kg	Reduce	LCA	Collect data over the first year of implementation for direct emissions (if significant). Effects further up supply chain more difficult to determine.
EV2: b) Impact on air (Nox)	kg	Reduce	LCA	As above
EV2: c) Impact on air (SO_4)	kg	Reduce	LCA	As above
EV3: CSO Discharge	m^3/yr	Reduce	Computer model simulation	Flow monitoring over minimum of 1 yr

* Knowledge is limited about the transport processes for SW in sewers.

REFERENCES

References cited in the Guidebook are presented here on a chapter by chapter basis. However, the first part of this section gives a list of useful or Key Texts that are considered important background to the development and/or use of this Guidebook. References cited in each of the case studies are presented at the end of each relevant appendix.

Key Texts

Alegre H., Hirner W., Baptista J M. and Parena R. (2000), Performance indicators for water supply services. IWA Manual of Best Practice, ISBN 1 900222 27 2.

Ashley R M., Alegre H., Matos R., Cardoso A., Duarte P., Molinari A. and Schulz A. (2002), The IWA Performance Indicator Manuals For Water And Wastewater Service Provision. Proc. PRO4 conference, Birmingham. July 9-11. Copyright IIR Ltd.

Chapman D. (Ed.) (1996), Water quality assessments. 2nd edition, Spon, ISBN 0 419 21590 5.

Cosgrove, W.J. and Rijsberman, F.R. for the World Water Council (2000), World Water Vision: Making Water Everybody's Business. Earthscan, London, ISBN 85383 730 X.

Cross, N. (1994), Engineering design methods: strategies for product design, 2nd Edition. Wiley, Chichester, ISBN 0471942286.

Department of the Environment, Transport and the Regions (1999), A better quality of life: a strategy for sustainable development in the UK. DETR, London, January.

Department of the Environment, Transport and the Regions (1999), Quality of life counts: Indicators for a strategy for sustainable development for the UK. DETR, London, December.

Department of the Environment, Transport and the Regions (2000), Local Quality of Life Counts: A handbook for a menu of local indicators of sustainable development. DETR, London, May.

Department of the Environment, Transport and the Regions (2000), Multi-Criteria Analysis: A Manual. DETR, London, December.

Ding, G in Best, R. and De Valence, G. (1999), Building in value. London: Arnold, ISBN 0 340 74160 0.

Environment Agency (1998), Best Practicable Environmental Option Assessments for Integrated Pollution Control. HMSO. ISBN 0 11 310126 0. 3rd impression.

European Commission (2001), Consultation paper for the preparation of a European Union strategy for Sustainable Development. Sustainable Development Task Force, Brussels.

European Commission (2000), Proposal for a Directive of the European Parliament and of the Council providing for public participation in respect of the drawing up of certain plans and programmes relating to the environment and amending Council Directives 85/337/EEC and 96/61/EC. Brussels, January, 2000/0331 (COD).

European Commission (2000), Directive of the European Parliament and of the Council 2000/60/EC Establishing a Framework for Community Action in the Field of Water Policy. Luxembourg, 23 October.

Everard, M. and Street, P. (2002), Putting Sustainable Drainage Systems (SUDS) into Practice. 2020 Vision Series No 7. The Natural Step, August.

Foundation for Water Research (FWR) (1998), Urban Pollution Management Manual R/CL0009. 2nd Edition (CD version).

Lens P., Zeeman G. and Lettinga G. (Eds.) (2001), Decentralised Sanitation and reuse – concepts, systems and implementation. IWA publishing. ISBN 1 900222 47 7.

Loucks, D. P. and Gladwell, J. S. (Eds.) (1999), Sustainability criteria for water resource systems. Cambridge University Press, ISBN 0 521 56044 6.

Maksimovic, C. and Tejada-Guibert, J.A. (Eds.) (2001), Frontiers in Urban Water Management. Deadlock or hope. IWA Publishing, London, ISBN 1 900222 76 0.

Mangione, T.W. (1995), Mail Surveys: Improving the Quality, London: Sage, ISBN 0 8039 4662 7 (hbk) and 0 8039 4663 5 (pbk).

Merret, S. (1997), Introduction to the economics of water resources – an international perspective, UCL Press, London, ISBN 1 85728 636 7 HB and 1 85728 637 5 PB.

Norton, B. R. and McElligott, W. C. (1995), Value management – a practical guide, MacMillan, London, ISBN 0 333 60626 4.

Pollard, S. and Guy, J. (2001). (Eds.), Risk Assessment for Environmental Professionals. Chartered Institution of Water and Environmental Management. December. ISBN 1 870752 71 6.

Roy, B. (1996), Multicriteria methodology for Decision Aiding. Kluwer Academic Press, The Netherlands, ISBN 0 7923 4166 X.

Starkey, C.V. (1992), Engineering design decisions. Arnold, London, ISBN 0 340 54378 7.

Stewart, D. W and Shamdasani, P. N. (1990), Focus groups: theory and practice, Sage, London, ISBN 0 8039 3389 4 (hbk) and 0 8039 3390 8 (pbk)

Strang, V. (2001), Evaluating Water – Cultural beliefs and values about Water Quality, Use and Conservation. Water UK Publications. ISBN 1 84057 248 5.

UKWIR (2001), Simplified Post Project Appraisal of CSO Performance: report reference WW/08/H.

UN WEHAB Working Group (2002), A Framework for Action on Water and Sanitation: World Summit on Sustainable Development, Johannesburg, August.

Vincke, P. (1992), Multicriteria Decision-Aid. John Wiley and Sons, ISBN 0 471 93184 5.

Walker, R. (1985), Applied qualitative research, Gower Publishing, Aldershot, ISBN 0 566 00897 I (cased) & 0 566 00898 X (paper).

Water UK (2000), Towards environmental sustainability: indicators for the UK water industry. May.

Wrisberg, N., Udo de Haes, H. A., Triebswetter, U., Eder, P. and Clift, R. (Eds.) (2002), Analytical Tools for Environmental Design and Management in a Systems Perspective: The Combined Use of Analytical Tools. Dordrecht, The Netherlands: Kluwer Academic Press, ISBN 1 4020 0453 2.

Preface References

European Commission (2000), Directive of the European Parliament and of the Council 2000/60/EC Establishing a Framework for Community Action in the Field of Water Policy. Luxembourg, 23 October.

IUCN/UNEP/WWF (1991), Caring for the Earth: A Strategy for Sustainable Living, Earthscan, London, ISBN 1 85383 126 3.

World Commission on Environment and Development (1987), Our Common Future ('The Brundtland report'), Oxford University Press, Oxford, ISBN 0 19 282080 X.

References for Chapter 1

ASCE and UNESCO (1998), Sustainability criteria for water resource systems, prepared by the task committee on sustainability criteria, American Society of Civil Engineers, and the working group of UNESCO/IHP IV Project M-4.3.

Balkema, A. (2003) Sustaianable Wastewater Treatments. PhD Thesis, University of Eindhoven. ISBN 90 386 1805 0.

Balkema A.J., Preisig H A., Otterpohl R., Lambert A J D., Weijers S R. (2001), Developing a model based decision support tool for the identification of sustainable treatment options for domestic wastewater, Water Science and Technology Vol. 43, No.7, pp. 265-269.

Balkema, A., Weijers, S.R. and Lambert, F.J.D. (1998), On methodologies for comparison of wastewater treatment systems with respect to sustainability, Proc. WIMEK Congress on Options for Closed Water Systems, March 11-13, Wageningen, The Netherlands.

Bentham, J. (1789), An Introduction to the Principles of Morals and Legislation. ed. by J.H. Burns, H.L.A. Hart and F. Rosen, Clarendon Press (March 1996), ISBN 0 19 820516 3.

World Commission on Environment and Development (1987), Our Common Future ('The Brundtland report'), Oxford University Press, Oxford, ISBN 0 19 282080 X.

Butler, D. and Parkinson, J. (1997), Towards sustainable urban drainage. Water Science and Technology 35(9), pp. 53–63.

Department of the Environment, Transport and the Regions (DETR) (1999a), A Better Quality of Life: A strategy for sustainable development for the UK, DETR, London, January.

Department of the Environment, Transport and the Regions (DETR) (1999b), Quality of Life Counts: Indicators for a Strategy for Sustainable Development for the UK, DETR, London, December.

European Commission (EC) (2000), Water Framework Directive 2000/60/EC.

Foxon, T.J., McIlkenny, G., Gilmour, D., Oltean-Dumbrava, C., Souter, N., Ashley, R., Butler, D., Pearson, P., Jowitt, P., Moir, J. (2002) Sustainability Criteria for Decision Support in the UK Water Industry. Journal of Environmental Planning and Management, 45(2), pp. 285-301.

Graaf, J.H.J.M. van der, Meester-Broertjes, H.A., Bruggeman, W.A. and Vles, E.J. (1997), Sustainable technological development for urban-water cycles, Water Science and Technology 35(10), pp. 213-220.

Hellstrom, D., Jeppsson, U. and Karrman, E. (1999), Systems analysis of sustainable water management - a first approach, Proc. 4th International Conference - Managing the Wastewater Resource, June 7-11, 1999, As, Norway.

Hellstrom, D., Jeppsson, U. and Karrman, E. (2000), A framework for systems analysis of urban water management, Environmental Impact Assessment Review 20(3), pp. 311-321.

Heywood, G., Lumbers, J., Reid, S., Ballance, T., Chalmers, L. and Haywood-Smith, B. (2002) Capital Maintenance Planning: A Common Framework. VOls 1-4. UK Water Industry Research Ltd (UKWIR).Report No. 02/RG/05/3. ISBN 1 84057 265 5.

IUCN/UNEP/WWF (1991), Caring for the Earth: A Strategy for Sustainable Living, Earthscan, London, ISBN 1 85383 126 3.

Jeppsson, U., Hellstrom, D. and Karrman, E. (1998), Systems analysis of sustainable water management, MISTRA project report, October 1998.

Larsen T.A. and Gujer W. (1997), Sustainable urban water management – technological implications. EAWAG News 43E, December, pp. 12-14.

Loucks, D.P. and Gladwell, J.S. (Eds.)(1998), Sustainability criteria for water resource systems, International Hydrology Series, Cambridge University Press, Cambridge, ISBN 0521 56044 6.

Lundin, M. (1999), Assessment of the environmental sustainability of urban water systems, Licentiate Thesis, Dept. of Technical Environmental Planning, Chalmers University of Technology, Goteborg, Sweden, 1999.

Lundin, M. (2003), Indicators for measuring the sustainability of urban water systems: A life cycle approach. PhD Thesis, Chalmers University of Technology. ISBN 91 7291 251 0.

Lundin, M., Molander, S. and Morrison, G.M. (1999), A set of indicators for the assessment of temporal variations in the sustainability of sanitary systems, Water Science and Technology, 39(5), pp. 235-242.

Matos, R., Cardoso, A., Ashley, R.M., Duark, O., Molinari, A. and Schulz, A. (2003), Performance Indicators for Wastewater Services. IWA Manual of Best Practice. ISBN 1 900222 906.

Niemczynowicz, J. (1994), New aspects of urban drainage and pollution reduction towards sustainability. Water Science and Technology, 30(5), pp. 269-277.

Penning-Rowsell, E., Johnson, C., Tunstall, S., Tapsell, S., Morris, J., Chatterton, J., Coker, A. and Green, C. (2003) The benefits of flood and coastal defence: techniques and data for 2003. Flood Hazard Research Centre, Middlesex University. Available from fhrcl@mdx.ac.uk

Pinchot, G. (1947), Breaking new ground. Harcourt, Brace, and Co., New York. Republished by Island Press (1 December 1987), ISBN 0 9332 8042 4.

Robert K-H., Daly H., Hawken P., Holmberg J. (1997), A compass for sustainable development. International Journal of Sustainable Development and World Ecology 4, pp. 79-92.

Skipworth, P., Engelhardt, M., Cashman, A., Savic, D., Saul, A. and Walters, G. (2002) Whole life costing for water distribution network management, Thomas Telford Publishing, London, ISBN 0 7277 31661.

Strang, V. (2001), Evaluating water: cultural beliefs and values about water quality, use and conservation. Water UK, ISBN 1 84057 248 5.

Upham P. (2000), Scientific consensus on sustainability: the case of the natural step. Sustainable development 8, pp. 180 – 190

Vachos, E. and Braga, B. (2001), The challenge of urban water management. Chapter 1 in Frontiers of Urban Water Management. Deadlock or Hope? (C. Maksimovic and J. Tejada-Guibert, eds.), IWA Publishing, London, ISBN 1 9002 2276 0.

Water UK (2000), Towards Environmental Sustainability: UK Water Industry Environmental Sustainability Indicators, Water UK, London.

Water UK (2003), Source 2003: Perspectives on Global Sustainability. Water UK in association with Ernst and Young. Available from Water UK.

Whelan, A. (2003), Water UK Investor Key Findings. April. Available from Water UK.

References for Chapter 2

Alegre H., Hirner W., Baptista J M. and Parena R. (2000), Performance indicators for water supply services. IWA Manual of Best Practice, ISBN 1 900222 27 2.

Ashley R M., Alegre H., Matos R., Cardoso A., Duarte P., Molinari A. and Schulz A. (2002), The IWA Performance Indicator Manuals For Water And Wastewater Service Provision. Proc. PRO4 conference, Birmingham. July 9-11. Copyright IIR Ltd.

Ashley R.M., Smith H., Jowitt P., Butler D., Blackwood DJ., Davies J., Gilmour D., Foxon T. (2002), Making more sustainable decisions for asset investment in the water industry – Sustainable Water industry Asset Resource Decisions – the SWARD project. 9th International Conference on Urban Drainage, Portland USA, September.

Astley, WG, Axelsson R., Butler, R.L., Hickson, D.J. and Wilson, D C. (1982), Complexity and cleavage; dual explanations of strategic decision making. Journal of Management Studies 19(4), pp. 357-375.
Bouchart F.J-C., Blackwood D.J. and Jowitt P.W. (2002), Decision Mapping: Understanding decision making processes. Civil Engineering and Environmental Systems, 19(3), pp. 187-207.
Brans, J.P., Mareschal, B., and Vincke, P. (1984), PROMETHEE: A new family of outranking methods in multicriteria analysis. In: Brans JP. (Ed.). Operational Research '84. Elsevier, North Holland, Amsterdam, pp. 408-421.
Cross, N. (1989), Engineering design methods, Chichester: Wiley. ISBN 0 471 92215 3.
Department of the Environment, Transport and the Regions (2000), Multi-Criteria Analysis: A Manual. DETR, London, December.
Edwards, W. and Newman, J.R. (1982), Multiattribute Evaluation. Sage, London, ISBN 0 8039 0095 3.
Keeney, R.L., Raiffa, H. and Meyer, R. (1993), Decisions with Multiple Objectives, Preferences and Value Tradeoffs. Cambridge University Press, Cambridge, ISBN 0 5214 3883 7.
Lindblom, CE. (1959), The science of muddling through, Public Administration Review, 19, pp. 72-88.
March JG. and Simon HA. (1958), Organisations, Wiley, New York, ISBN 0 63118631 X.
MORI (2002), The 2004 periodic review: research into customers' views, November.
Mylius A. (2002), Anglian challenges EU water treatment directive. New Civil Engineer, 21, March.
Norton, B. R. and McElligott, W. C. (1995), Value management – a practical guide, MacMillan, London, ISBN 0 333 60626 4.
Rodgers M. (2001), Engineering project appraisal, Blackwell Science, Oxford, ISBN 0 6320 5606 1.
Roy, B. (1991), The outranking approach and the foundations of ELECTRE methods. Theory and Decision 31(1), pp. 49-73.
Saaty, T.L. (1980), The Analytic Hierarchy Process. McGraw-Hill, New York, ISBN 0 0705 4371 2.
Simon H.A. (1997), Administrative behaviour, Simon & Schuster, ISBN 0 6848 3582 7.

Simon H.A. (1984), The Structure of ill-structured problems, in Developments in design methodology, Ed Cross N, Wiley, Chichester, ISBN: 0 4711 0248 2.
Sustainability Assessment Model (SAM) used by BP. For more information access: http://www.abdn.ac.uk/oilgas/acc.htm
UKWIR (2002), Capital Maintenance Planning: A Common Framework. UKWIR.
Wrisberg, N., Udo de Haes, H. A., Triebswetter, U., Eder, P. and Clift, R. (Eds.) (2002), Analytical Tools for Environmental Design and Management in a Systems Perspective: The Combined Use of Analytical Tools. Dordrecht, The Netherlands: Kluwer Academic Press, ISBN 1 4020 0453 2.

References for Chapter 3

Byatt I. C. R. (2000), Maintaining Serviceability to customers. Letter to Managing Directors of WSPs. MD161. April.
Chadwick A. (2002), Socio-economic impacts: are they still the poor relations in UK environmental statements? Journal of Environmental Planning and Management, 45(1), pp. 3-24.
Chapman D. (Ed.) (1996), Water quality assessments. 2nd edition, Spon, London, ISBN 0 419 21590 5.
Cross, N. (1994), Engineering design methods: strategies for product design, 2nd Edition. Wiley, Chichester, ISBN 0471942286.
Ding, G in Best, R. and De Valence, G. (1999), Building in value. Arnold, London, ISBN 0 340 74160 0.
Foxon, T J, Leach, M, Butler, D, Dawes, J, Hutchinson, D, Pearson, P and Rose, D (1999), Useful indicators of urban sustainability: some methodological questions, Local Environment, 4(2), (June 1999), pp. 137-149
Foxon, TJ., Butler, D., Dawes, J., Hutchinson, D., Leach, M., Pearson, P. and Rose, D. (2000), 'An assessment of water demand management options from a systems approach', Journal of Chartered Institute of Water and Environmental Management 14, pp. 171-178.
Heywood, G., Lumbers, J., Reid, S., Ballance, T., Chalmers, L. and Haywood-Smith, B. (2002), Capital Maintenance Planning: A Common Framework. VOls 1-4. UK Water Industry Research Ltd (UKWIR).Report No. 02/RG/05/3. ISBN 1 84057 265 5.
Leach, M, Bauen, A and Lucas, N (1997), A systems approach to material flows in sustainable cities: Case study of paper, Journal of Environmental Planning and Management, 40(6), pp. 705-723
Mangione, T.W. (1995), Mail Surveys: Improving the Quality, Sage, London, ISBN 0 8039 4662 7 (hbk) and 0 8039 4663 5 (pbk).

Meier, P (1984), Energy Systems Analysis for Developing Countries, Lecture Notes in Economics and Statistics, 222, Springer-Verlag, Berlin

Merret, S. (1997), Introduction to the economics of water resources – an international perspective, UCL Press, London, ISBN 1 85728 636 7 HB and 1 85728 637 5 PB.

Penning-Rowsell, E., Johnson, C., Tunstall, S., Tapsell, S., Morris, J., Chatterton, J., Coker, A. and Green, C. (2003), The benefits of flood and coastal defence: techniques and data for 2003. Flood Hazard Research Centre, Middlesex University. Available from fhrcl@mdx.ac.uk

PRé consultants B.V. (1997), SimaPro Single user database manual.

Skipworth, P., Engelhardt, M., Cashman, A., Savic, D., Saul, A. and Walters, G. (2002) Whole life costing for water distribution network management, Thomas Telford Publishing, London, ISBN 0 7277 31661.

Starkey, C.V. (1992), Engineering design decisions. Arnold, London, ISBN 0 3405 4378 7.

Stewart, D. W and Shamdasani, P. N. (1990), Focus groups: theory and practice, Sage, London, ISBN 0 8039 3389 4 (hbk) and 0 8039 3390 8 (pbk)

Strang V. (2001), Evaluating Water – Cultural beliefs and values about Water Quality, Use and Conservation. Water UK Publications. ISBN 1 84057 248 5.

UKWIR (2001), Simplified Post Project Appraisal of CSO Performance: report reference WW/08/H.

Mistra (2001), Urban Water – Progress Report 2001. Chalmers Tekniska Hogskola. ISSN 1650 3791.

Walker, R. (1985), Applied qualitative research, Gower Publishing, Aldershot, ISBN 0 566 00897 I (cased) & 0 566 00898 X (paper).

Water UK (2000), Towards environmental sustainability: indicators for the UK water industry. May.

Wrisberg, N., Udo de Haes, H. A., Triebswetter, U., Eder, P. and Clift, R. (Eds.) (2002), Analytical Tools for Environmental Design and Management in a Systems Perspective: The Combined Use of Analytical Tools. Kluwer Academic Press, Dordrecht, The Netherlands, ISBN 1 4020 0453 2.

APPENDICES

Appendix A: Sustainability Indicators

Contents

UK Government

In 1999 the UK Government published 'A better quality of life: a strategy for sustainable development in the UK'. This report first presented the UK Government's Headline Indicators for Sustainable Development (DEFRA, 1999a, 1999b). These 15 indicators are intended to raise public awareness of Sustainable Development, to focus public attention on what Sustainable Development means, and to give a broad overview of progress. The indicators cover the three pillars of Sustainable Development: Economic growth, Social progress and Environmental protection. **Table A1** illustrates how the Headline Indicators fit into the Government's overall strategy for Sustainable Development. A set of around 150 core indicators are further used to monitor national progress. These core indicators give a broad overview of whether the UK is achieving a 'better quality of life for everyone, now and for future generations to come'. They cover the three pillars of Sustainable Development, yet include people's everyday concerns - health, jobs, crime, air quality, traffic, housing, educational achievement, wildlife and economic prosperity. Tables **A2a-A2u** present the full list of the UK Government's Sustainable Development objectives and indicators.

Further information about the UK Government's Sustainable Development strategy can be accessed at http://www.sustainable-development.gov.uk/

Water UK

The Water UK environmental sustainability indicators, launched in 2000, are classified under the following subgroups:

A Water services
B Good environmental management
C Biodiversity and the environment
D Energy and materials

These environmental indicators are presented in **Table A3**. The data used to derive the indicators in Group A are already collected for regulatory purposes. These data have usually been independently verified and their confidence rating is high. Other data are gathered for operational and management purposes or for the specific purpose of establishing environmental sustainability indicators. Water UK are currently working with stakeholders to improve the environmental indicators and are to expand the number of indicators to 37, to include economic and social considerations. The revised set of indicators will be grouped into six new categories:

A Governance, strategic planning and management
B Society
C Environment
D Finance
E Assets
F Employees

Reporting for the new set is due to start in 2004. The indicators will provide a more comprehensive picture of how the UK water industry as a whole is moving towards sustainability. Further information about the Water UK Sustainability indicators can be accessed via http://www.water.org.uk/index.php

Table A1. The headline indicators in the UK sustainable development strategy

Themes, issues and objectives	Ref.	Headline Indicators
Maintaining high and stable levels of economic growth and employment		
our economy must continue to grow	H1	GDP and GDP per head
investment (in modern plant and machinery as well as research and development) is vital to our future prosperity	H2	Total and social investment as a percentage of GDP
maintain high and stable levels of employment so everyone can share greater job opportunities	H3	Percentage of people of working age who are in work
Social progress which recognises the needs of everyone		
tackling poverty and social exclusion	H4	Indicators of success in tackling poverty and social exclusion (children in low income households, adults without qualifications and in workless households, elderly in fuel poverty)
equip people with the skills to fulfil their potential	H5	Qualifications at age 19
improve health of the population overall	H6	Expected years of healthy life
reduce the proportion of unfit (housing) stock	H7	Non-decent housing
reduce both crime and peoples fear of crime	H8	Level of crime
Effective protection of the environment		
continue to reduce our emissions of greenhouse gases now, and plan for greater reductions in longer term	H9	Emissions of greenhouse gases
reduce air pollution and ensure air quality continues to improve through the longer term	H10	Days when air pollution is moderate or higher
improve choice in transport; improve access to education, jobs leisure and services; and reduce the need to travel	H11	Road traffic
improving river quality	H12	Rivers of good or fair quality
reverse the long-term decline in populations of farmland and woodland birds	H13	Populations of wild birds
re-using previously developed land, in order to protect the countryside and encourage urban regeneration	H14	New homes built on previously developed land
Prudent use of natural resources		
move away from disposal of waste towards waste minimisation, reuse, recycling and recovery	H15	Waste arisings and management

*Table A2a. Complete list of UK Government objectives and indicators: Sustainable Economy (**D** = Indicator to be developed)*

Objective	Ref	Indicators
A SUSTAINABLE ECONOMY		
A Doing more or less: improving resource efficiency		
Greater resource efficiency	A1	UK resource use D
Energy efficiency of the economy	A2	Energy efficiency of economy
	A3	Energy use per household
Move away from disposal of waste towards waste reduction, reuse, recycling and recovery	H15	Waste arisings and management
	A4	Waste by sector D
	A5	Household waste and recycling
	A6	Materials recycling
	A7	Hazardous waste

Table A2b. Complete list of UK Government objectives and indicators: Economic stability and competitiveness

Objective	Ref	Indicators
B Economic stability and competitiveness		
Our economy must continue to grow	H1	Total output of the economy (GDP and GDP per head)
Deliver low inflation	B1	Rate of inflation
Government borrowing at stable and prudent level	B2	Public sector net borrowing and net debt
Promote UK competitiveness	B3	Labour productivity
	B4	UK imports, exports, trade balance
Investment (in modern plant and machinery, as well as research and development) is vital to our future prosperity	H2	Total and social investment as a percentage of GDP
Also invest in "social assets"	B5	Social investment as a per cent of GDP

*Table A2c. Complete list of UK Government objectives and indicators: Developing skills and rewarding work (**D** = Indicator to be developed)*

Objective	Ref	Indicators
C Developing skills and rewarding work		
Equip people with the skills to fulfil their potential	H5	Qualifications at age 19
Raise educational standards at all levels and close the widening gap between high and low achievers	C1	16 year-olds with no qualifications
	C2	Adult literacy/numeracy
To become a learning society - in a rapidly changing world people need the skills to adapt, and opportunities to update them throughout their lives	C3	Learning participation
Boost workplace learning	C4	Businesses recognised as Investors In People
Maintain high and stable levels of employment so everyone can share greater job opportunities	H3	Proportion of people of working age who are in work
	C5	Proportion of people of working age in workless households
	C6	Proportion of people of working age out of work for more than two years
	C7	Proportion of lone parents, long-term ill and disabled people who are economically active
Fairness at work	C8	People in employment working long hours
	C9	Low pay D
Maintain a safe and healthy environment for workers	C10	Work fatalities and injury rates; working days lost through illness
Raise quality of life of workers in global supply chains of companies importing into the UK	C11	UK companies implementing ethical trading codes of conduct D

*Table A2d. Complete list of UK Government objectives and indicators: Sustainable production and consumption (**D** = Indicator to be developed)*

Objective	Ref	Indicators
D Sustainable production and consumption		
Give consumers better information and encourage purchasing initiatives which help to move the market	D1	Consumer information D
Contextual indicator	D2	Consumer expenditure
Take-up of best practice in key sectors	D3	Energy and water consumption by sector/Waste and hazardous emissions by sector D
Encourage businesses to assess environmental impacts and set targets, and produce environmental reports	D4	Adoption of environmental management systems (ISO 14001) and the EU Eco-Management and Audit Scheme (EMAS)
	D5	Corporate environmental engagement
	D6	Environmental reporting D
Need housing which is more energy-efficient, uses fewer resources and creates less waste	D7	Household water use and peak demand
	D8	Thermal efficiency of housing stock
Greater use of sustainable construction materials	D9	Primary aggregates per unit of construction value
	D10	Construction and demolition waste going to landfill
Need more efficient appliances	D11	Energy efficiency of new appliances
Minimise the impact of pesticides on human health	D12	Pesticide residues in food
Encourage environmentally sensitive land management by farmers	D13	Area under agreement under the Environmentally Sensitive Area and Countryside Stewardship agri-environment schemes
Organic production methods	D14	Area converted to organic production
Encourage production, marketing, purchase and use of vehicles that are more fuel efficient	D15	Energy efficiency of road passenger travel/Average fuel consumption of new cars
UK tourism industry to grow significantly in ways which are economically, socially and environmentally beneficial	D16	Sustainable tourism D
Address transport issues; integrate tourism with public transport	D17	Leisure trips by mode of transport
	D18	Overseas travel
Reduce environmental impact of chemicals	D19	Chemical releases to the environment D
Develop distribution systems which support economic growth, protect the environment and benefit society	D20	Freight transport by mode
	D21	Heavy goods vehicle mileage intensity

Table A2e. Complete list of UK Government objectives and indicators: Promoting economic vitality and employment

Objective	Ref	Indicators
BUILDING SUSTAINABLE COMMUNITIES		
E Promoting economic vitality and employment		
Improve economic performance and enhance regional competitiveness	E1	Regional variations in GDP
Closing the gap between the poorest communities and the rest	E2	Index of local deprivation
Tackling poverty and social exclusion	H4	Indicators of success in tackling poverty and social exclusion
	E3	Truancies and exclusions from school/teenage pregnancies
Promoting local business diversity	E4	New business start-ups net of closures
Reducing disproportionate unemployment among ethnic minorities	E5	Ethnic minority employment and unemployment

Table A2f. Complete list of UK Government objectives and indicators: Better health for all

Objective	Ref	Indicators
F Better health for all		
Improve health of the population overall	H6	Expected years of healthy life
Deliver key health targets	F1	Death rates from cancer, circulatory disease, accidents and suicides
Environmental factors affecting health	F2	Respiratory illness
Address major factors leading to health inequalities	F3	Health inequalities
Provide people with access to effective healthcare, based on patients' needs, and not on where they live or their ability to pay	F4	NHS hospital waiting lists

Table A2g. Complete list of UK Government objectives and indicators: Travel

Objective	Ref	Indicators
G Travel		
Improve choice in transport; improve access to education, jobs, leisure and services; and reduce the need to travel	H11	Road traffic
	G1	Passenger travel by mode
	G2	How children get to school
	G3	Average journey length by purpose
The cost of traffic congestion	G4	Traffic congestion
The link between rising prosperity and increased travel must be broken	G5	Distance travelled relative to income

Table A2j. Complete list of UK Government objectives and indicators: Access

Objective	Ref	Indicators
J Access		
Need better access to services	J1	People finding access difficult
	J2	Access to services in rural areas
Ensure that disabled people have access to a wider range of goods, services and facilities	J3	Access for disabled people
Arts and sport should be accessible to everyone	J4	Participation in sport and cultural activities
Reduce the proportion of unfit (housing) stock	H7	Homes judged unfit to live in
Ensure that everyone has the opportunity of a decent home	J5	Temporary accommodation/rough sleepers
Improving significantly the energy efficiency of all residential accommodation	J6	Fuel poverty

Table A2k. Complete list of UK Government objectives and indicators: Shaping our surroundings (D = Indicator to be developed)

Objective	Ref	Indicators
K Shaping our surroundings		
Re-using previously developed land, in order to protect the countryside and encourage urban regeneration	H14	New homes built on previously developed land
Bring empty homes back into use and convert buildings to new uses	K1	Vacant land and properties and derelict land
Shopping, leisure and entertainment, offices and other key town centre uses should, wherever possible, be located within existing centres	K2	New retail floor space in town centres and out of town
Contextual indicators	K3	Population growth
	K4	Household growth
Ensure that development takes account of history and look for opportunities to conserve local heritage	K5	Buildings of Grade I and II* at risk of decay
Attractive streets and buildings, low levels of traffic, noise and pollution, green spaces, and community safety	K6	Quality of surroundings
	K7	Access to local green space D
	K8	Noise levels
Reduce both crime and fear of crime	H8	Level of crime
	K9	Fear of crime

Table A2l. Complete list of UK Government objectives and indicators: Involvement and stronger institutions

Objective	Ref	Indicators
L Involvement and stronger institutions		
All local communities to have sustainable development strategies in place by 2000	L1	Number of local authorities with LA21 strategies
Voluntary and community activity can promote social inclusion and cohesion	L2	Voluntary activity
Help build a sense of community by encouraging and supporting all forms of community involvement	L3	Community spirit

Table A2m. Complete list of UK Government objectives and indicators: An integrated approach

Objective	Ref	Indicators
MANAGING THE ENVIRONMENT AND RESOURCES		
M An integrated approach		
Must not store up pollutant problems for the future	M1	Concentrations of persistent organic pollutants
	M2	Dangerous substances in water
	M3	Radioactive waste stocks
	M4	Discharges from the nuclear industry

Table A2n. Complete list of UK Government objectives and indicators: Climate change and energy supply

Objective	Ref	Indicators
N Climate change and energy supply		
Climate change must be kept within limits which global society can accommodate	N1	Rise in global temperature
Assess vulnerability to changed weather patterns and higher sea levels and develop adaptation strategies	N2	Sea level rise
Continue to reduce our emissions of greenhouse gases now, and plan for greater reductions in the longer term	H9	Emissions of greenhouse gases
	N3	Carbon dioxide emissions by end user
In the longer term more energy will have to come from new and renewable sources	N4	Electricity from renewable sources
Fossil fuel resources managed in an environmentally-acceptable way	N5	Depletion of fossil fuels

Table A2p. Complete list of UK Government objectives and indicators: Air and atmosphere

Objective	Ref	Indicators
P Air and atmosphere		
Reduce air pollution and ensure air quality continues to improve through the longer term	H10	Days when air pollution is moderate or higher
	P1	Concentrations of selected air pollutants
	P2	Emissions of selected air pollutants
Ensure that polluting emissions do not cause harm to human health or the environment	P3	Sulphur dioxide and nitrogen oxides emissions
	P4	Acidification in the UK
Controls on ozone depleting substances	P5	Ozone depletion

Table A2q. Complete list of UK Government objectives and indicators: Freshwater (D = Indicator to be developed)

Objective	Ref	Indicators
Q Freshwater		
Improving river quality	H12	Rivers of good or fair quality
	Q1	Nutrients in water
Safeguarding resources and ensuring affordable supplies	Q2	Water demand and availability
	Q3	Water affordability
Avoiding waste of water	Q4	Water leakage
	Q5	Abstractions by purpose
Ensure that abstraction controls play a full part in protecting the best wildlife and amenity sites	Q6	Sites affected by water abstraction D

Table A2r. Complete list of UK Government objectives and indicators: Seas oceans and coasts (D = Indicator to be developed)

Objective	Ref	Indicators
R Seas oceans and coasts		
Reduce or eliminate inputs of hazardous and radioactive substances of most concern	R1	Estuarine water quality, marine inputs
Aim to raise consistent compliance with the European Bathing Water Directive	R2	Compliance with Bathing Water Directive
Protection of marine habitats and species	R3	Biodiversity in coastal/marine areas D
Improve the management and conservation of fish stocks	R4	Fish stocks around the UK fished within safe limits
Work with other countries to achieve effective management and conservation of fish stocks	R5	State of the world's fisheries

Table A2s. Complete list of UK Government objectives and indicators: Landscape and wildlife (D = Indicator to be developed)

Objective	Ref	Indicators
S Landscape and wildlife		
Minimise the loss of soils to new development	S1	Net loss of soils to development
Soil protection	S2	Concentrations of organic matter in agricultural topsoils
Reverse the long-term decline in populations of farmland and woodland birds	H13	Populations of wild birds
	S3	Trends in plant diversity
Reverse the decline in UK wildlife and habitats	S4	Biodiversity action plans
Protection for individual landscape features such as hedges, dry stone walls and ponds	S5	Landscape features - hedges, stone walls and ponds
Strengthen protection for special sites	S6	Extent and management of SSSIs
Protecting the wider landscape	S7	Countryside quality D
Promoting public access and enjoyment of the landscape	S8	Access to the countryside D
Contextual indicator	S9	Native species at risk
Continuing expansion of UK woodland area	S10	Area of woodland in the UK
Protecting and expanding ancient and semi-natural woodlands	S11	Area of ancient semi-natural woodland in GB
Better management of existing woodlands	S12	Sustainable management of woodland D
Sustainable forestry management overseas	S13	Number of countries with national forest programmes
Aim to maximise efficient use of materials and greater use of recycled and waste materials	S14	Amount of secondary/ recycled aggregates used compared with virgin aggregates D
All mineral working sites are restored to a standard suitable for a specific beneficial afteruse	S15	Land covered by restoration and aftercare conditions

Table A2t. Complete list of UK Government objectives and indicators: Sending the right signals

Objective	Ref	Indicators
SENDING THE RIGHT SIGNALS		
T Sending the right signals		
Integrating the environment into each department's policies and operations	T1	Greening government operations
Help promote women's interests and fairness at work	T2	Women in public appointments and senior positions
Contextual indicator	T3	Prices of key resources - fuel
Explore the scope for using economic instruments to deliver more sustainable development, and avoid "perverse subsidies" which work against sustainable development	T4	Real changes in the cost of transport
Cost-effective ways to comply with pollution abatement and aim to move to cleaner processes in the long term	T5	Expenditure on pollution abatement
Where new regulation is used it will conform to the government's principles of better regulation and be enforced effectively	T6	Enforcement of regulations
Improve awareness of sustainable development	T7	Public understanding and awareness
	T8	Awareness in schools
Encourage individuals to do their bit	T9	Individual action for sustainable development

Table A2u. Complete list of UK Government objectives and indicators: International co-operation and development (D = Indicator to be developed)

Objective	Ref	Indicators
INTERNATIONAL CO-OPERATION AND DEVELOPMENT		
U International co-operation and development		
Work with others to eliminate global poverty and raise living standards in developing countries	U1	Global poverty
	U2	Net Official Development Assistance (oda)
Contextual indicator	U3	Global population
Work with others to tackle global pressures on the environment and resources	U4	UK public expenditure on global environment protection
International conventions aimed at protecting the environment: UK commitment to Multilateral Environmental Agreements	U5	Implementation of multilateral environmental agreements D
Contextual indicators	U6	International emissions of carbon dioxide per head
	U7	World and UK materials consumption levels per head

Table A3. Water UK's 20 Environmental Sustainability Indicators

Subject	Indicator
Water Services	
A1 Water demand and availability	Population with sufficient resources (%)
	UK population growth possible with current resources (%)
A2 Household water demand	Per capita water consumption (litres per head per day)
A3 Non-household water use	Water efficiency (litres per £GDP)
A4 Leakage	Total leakage from the network (Megalitres per day)
A5 Drinking water quality	Tests complying with standards (%)
A6 Sewer flooding	Properties flooded (%)
A7 Intermittent discharges	Discharges in satisfactory condition (%)
A8 Wastewater treatment works	Population served by works meeting numerical standards (%)
Good Environmental Management	
B1 Environmental engagement	BiE Sectoral ranking (%)
B2 Convictions for public health and environmental offences	Number of category 1 convictions
Biodiversity and the Environment	
C1 Species	Priority species with action plans (%)
C2 Habitats	Priority habitats with action plans (%)
C3 River water quality	Rivers in classes A-D (%)
C4 Bathing water quality	Designated waters achieving mandatory standards (%)
	Designated waters achieving guideline values (%)
Energy and Materials	
D1 Energy use at fixed sites	Energy use per Ml water supplied (kWh)
	Energy used per Ml wastewater treated (kWh)
D2 Renewable energy at fixed sites	Renewable energy as percentage of total energy used (%)
D3 CO_2 emissions at fixed sites	Emissions per head population (tonnes/year)
D4 CO_2 emissions from road transport	Emissions per head population (tonnes/year)
D5 Sludge management	Sludge recycled/reused (%)
D6 Excavated materials	Materials recycled (%)

REFERENCES

Department of the Environment, Transport and the Regions (DETR) (1999a), A Better Quality of Life: A strategy for sustainable development for the UK, DETR, London, January.

Department of the Environment, Transport and the Regions (DETR) (1999b), Quality of Life Counts: Indicators for a Strategy for Sustainable Development for the UK, DETR, London, December.

Water UK (2000), Towards environmental sustainability: indicators for the UK water industry. Water UK, May.

Appendix B: Data Collection and Information Generation

Contents

INTRODUCTION

The option selection phase of the SWARD process requires that each option be assessed according to a number of sustainability indicators. The indicators help decision makers judge the relative performance of options according to social, environmental, economic and technical criteria. Options may be assessed by informal comparison of indicators or, more likely, by some sort of formalised multi-criteria analysis. Regardless of the method used, there is a need to collect data and generate information so that we can see how the options compare. But what do we mean by data and information?

In this context we define information as data that can be readily understood and therefore used in the decision making process. The SWARD sustainability indicators provide information on which to help assess relative sustainability. However, some of this information will be based on data that, by itself, may not be readily understood by all stakeholders in the decision making process.

Take, for example, the SWARD environmental sustainability indicator of "Energy use for water supply (kWh/cu m)". Generating this information (for potential projects) may require the collection of data about process types, machinery used and its energy consumption, control systems, operating patterns, volumes to be treated, treatment levels, raw water quality etc. Although the SWARD indicator is easily understood, the data behind it may not be.

Other sustainability indicators are more straightforward, and can make direct use of raw data. The resource utilisation indicator "Land area used (sq km)" is one example.

Thus, the assessment of options using sustainability indicators will require both data collection and information generation. In practice, these methods are closely linked: to generate information it will almost certainly be necessary to collect data. Bearing this in mind, a number of data collection/information generation methods will be discussed.

B1 DATA COLLECTION

Each WSP should have a clear idea of how the data is to be used before it is actually collected. The advice of Woodhouse (1999), discussing data collection for cost/risk optimisation, is relevant here. "Without knowing how we would use it, how do we know what data is worth collecting in the first place? Even if we are lucky enough to guess correctly on the data that is needed, how (and when) would we know that we had collected enough? What is enough, and is it physically/economically possible to collect it? Without a clear understanding of how it will be used, and the sensitivity to data inaccuracy, it is impossible to say what data is needed and to what precision."

Assuming that the "what", "how much", and "how precise" questions can be answered, the next question is "how to collect the data?"

B1.1 Records

It is possible that some of the data required (for sustainability indicators) will already have been recorded and will be readily accessible via the WSP's IT systems. If not then the WSP will have to make a special effort to collect it. Changes to the WSPs internal procedures may provide a means of obtaining the required data.

Maintenance staff, for example, are often required to record job details on a work order form. If a WSP needs more detailed data about reliability (e.g. for the risk of failure sustainability indicator) it could alter the work order form to include additional data entry fields. Work Type could be one such field, where staff may classify jobs as unplanned repairs, planned repairs, routine tasks or improvement work.

The most important factor in the design of such forms is simplicity. Forms should be as easy to complete as possible. Few people like filling out forms and the easier you can make it for them, the more likely you are to get the data you want. If possible, forms should record facts, not ask for subjective judgements.

Forms which are electronic i.e. have to completed directly via a computer terminal are preferable to paper based forms (although paper forms are often used to record job details temporarily until they are entered into the computer system). The data entry fields of computer based forms can usually be assigned a 'required' status. The form cannot then be signed off until these fields have been completed. Errors can be reduced (and form completion made easier) by the use of 'drop-down lists' that requires the user to pick from a number of options. Such forms, if well designed, can be completed in a few minutes and result in a negligible extra burden on staff.

B1.2 Direct Measurement

Some indicator data may be collected by direct measurement. Examples include the concentration of nutrients in a river (environmental impact on water) and the heavy metal content of sludge (the environmental impact on land).

B1.3 Collecting Data from Stakeholders

Stakeholders have already been defined in the guidebook as:

"Any individual or group which has a stake in how water and wastewater services are delivered, and could include customers, employees, environmental and community groups"

Data collected from employees (internal stakeholders) may be used by the WSP to help understand the effects of changes to water/wastewater systems. These effects could be economic (e.g. production costs), environmental (e.g. energy consumption), social (e.g. public access to watercourses) or technical (e.g. reliability). It is assumed that WSPs will already have effective ways of collecting data from their employees.

Data collected from customers, environmental groups and community groups (external stakeholders) is more likely to relate to the acceptability of the effects of changes to water/wastewater systems. An obvious example is the acceptability of something new being built in someone's backyard, be it a reservoir, treatment works, pumping station, or whatever. Less obvious examples could include the acceptability of domestic water meters, low flush toilets and grey water systems. Will changes be perceived as having a negative effect on the environment, health, etc.?

Various methods can be used to collect data from external stakeholders. East of Scotland Water, for example, makes use of customer focus groups; postal surveys; telephone surveys; community council meetings and public meetings. A "mobile information unit" is also used so that customers can "visit with queries about the water supply in their own area" (East of Scotland Water, 1999/2000)

We will briefly discuss data collection from external stakeholders by way of:

- Survey research;
- Focussed group interviews (focus groups);
- Individual 'depth' interviews.

There are many social research textbooks that describe these topics in greater detail.

B1.3.1 Survey research

McNeill (1990) describes a social survey as: "a method of obtaining large amounts of data, usually in a statistical form, from a large number of people in a relatively short time."

He goes on to suggest the typical stages of a survey:

1. Choosing the topic to be covered
2. Reviewing the literature
3. Forming of hunches and hypothesis
4. Identifying the population to be surveyed
5. Carrying out preparatory investigations and interviews
6. Drafting the questionnaire or interview schedule
7. Conducting a pilot survey
8. Finalising the questionnaire
9. Selecting a sample of the population
10. Selecting and training interviewers (if necessary)
11. Collecting the data
12. Processing the data and analysing the results
13. Writing the research report, perhaps in the form of a book
14. Publication of the report

We will describe the collection of survey data by use of questionnaires. These are lists of questions that are asked in exactly the same way of every respondent. They can be administered in a number of ways, including by post or by 'standardised interview' (face-to-face or telephone). Questionnaires used in standardised interviews are called 'interview schedules'. Each method has its advantages and disadvantages - these will be discussed later - but all are compatible with the generic survey stages outlined above.

The use of survey research will be illustrated by its theoretical application to the SWARD case study of Hall and Davies (**Appendix D5**). This is based on the redevelopment of an existing wastewater treatment works for which there are a number of options. The works, comprising two sites, are close to a national park and next to a tourist town, in full view of a main tourist route into the town. There are properties adjacent to both sites and residents can smell the works. A supermarket is also adjacent to one of the sites. Consequently, the project may have social, environmental or even economic effects (damage to businesses) that impact adversely on the local population. Let us suppose that the water company wishes to discover the likely social acceptability of each option before making a decision. It could conduct a social survey for this purpose, following the stages outlined earlier (note that stages 13 & 14 will be omitted, as they are about sharing findings with the social science community – not informing the decision making process of a single water company).

1. Choosing the topic to be covered

In this case, the topic is fairly clear: the acceptability of alternative options for the redevelopment of a waste water treatment works.

2. Reviewing the literature

It is possible that similar surveys, regarding the acceptability of potentially unpopular projects, have already been carried out and published. These may give ideas about research design, key issues and data collection methods as well as preventing the repetition of another's mistakes.

3. Forming of hunches and hypothesis

The researcher (who may be an employee of the water company or a contractor) should have some feeling for what the survey will reveal. They may, for example, believe that local residents will be most concerned about smell and noise problems and less concerned about the potential for employment at the works. They might also expect concern from environmental groups about effects on the national park. Furthermore, they may anticipate the manager of the supermarket (adjacent to one of the works' sites) to be significant source of disquiet.

Having made an educated guess about what to expect, the researcher will now be clearer about the questions that ought to be asked.

4. *Identification of the population to be surveyed*

The researcher must identify the population to be surveyed. This means "all those people who could be included" (Oppenheim, 1992). The researcher may include those people who live within a certain radius of the works. They may also include those people who own businesses within this radius (even if they live outside this radius). If the works predominantly affect a village or town, the researcher may use the boundary of this settlement to define the population.

5. *Carrying out preparatory interviews and questions*

Rather than depend entirely on their own hunches, a researcher may conduct informal interviews in order to find out what the key issues are (McNeill, 1990). The researcher may carry out such interviews with local residents, shopkeepers, and environmentalists to help discover what their concerns are about the redevelopment of the waste water treatment works. Mangione (1995) suggests that interviews may also be conducted with "knowledgeable observers of the issues you want to research". No doubt there are other water industry professionals who have faced similar problems in relation to the redevelopment of waste water treatment works. Mangione also suggests that focus groups, which will be described in more detail later, are effective in helping identify important issues.

6. *Drafting the questionnaire or interview schedule*

Question design is surprisingly complex, so much so that whole books have been written on the subject. However, Mangione condenses a great deal of relevant and readable advice into two chapters of the book "Mail Surveys: Improving the Quality". His 'guidelines for constructing questions' are summarised below.

- Write brief questions
  - They are more likely to be read completely
  - They are less likely to have qualifying phrases
  - They are less susceptible to biases produced by format/positioning of response alternatives
- Write clear questions
  - Define key terms (this should come at the beginning of the question)
  - Beware of expert jargon
  - Beware of unclear referents of pronouns (what does 'it' or 'they' refer to?)
  - Avoid double negatives
  - Avoid wording that starts with "how", "why", "where" or "how much" - at least for open-ended questions
- Stay in touch with reality
  - Remember that questions about people's intentions may not reflect actual future behaviour
  - Beware of hypothetical questions ("it depends" is a common reply)
  - Only ask relevant questions (allow people to skip irrelevant ones)
  - Commonplace is not universal (don't assume that people have only one job, house, etc)
- Avoid multi-dimensional questions
  - Questions should address one issue only (don't combine two or three questions within one question)
- Response categories should be mutually exclusive i.e.

  - Avoid age categories like 25-30, 30-35, 35-40 etc (to which category do 30,35 and 40 year olds belong?)
  - Avoid marital status categories such as married, single, widowed, divorced or separated (the single category overlaps with widowed, divorced and separated)
- Create exhaustive response categories
  - Consider the whole range of answers that someone might give
- Don't write loaded questions
  - Avoid bias that could force people to answer in one direction e.g. "most people think the local treatment works are too close. How do you feel?"

7. *Conducting a pilot survey*

This stage is necessary for a questionnaire-based survey, especially if it will be administered by post (the respondent has no way of querying the researcher). It is highly unlikely that the first draft of a questionnaire will be completely effective. Instead it is necessary to try it out, make improvements and then try it out again – often several times. This process can take anywhere from a few weeks to over a year. Although expensive, it avoids wasting effort on "unintelligible questions producing unquantifiable responses and uninterpretable results" (Oppenheim, P64).

Oppenheim (1992) recommends that "almost anything about a social survey can and should be piloted". The pilot survey should highlight problems such as poor question design, sequence and layout. The survey can also test administrative processes, sampling methods and the statistical treatment of results. Even the effects of different paper colours, typefaces, etc can be piloted.

The questionnaire should be tried out on a number of people who are similar to those of the main enquiry. Mangione suggests using groups of about 10 people to complete the questionnaire. A number of groups should be run so that a total of at least 25 respondents have been involved. After completing the questionnaire, a discussion should be held to gain overall feedback (survey length, appearance, ease of completion) as well as feedback on specific sections. Respondents should be encouraged to ask about things they don't understand and to contribute ideas for improvement.

8. *Finalising the questionnaire*

At this point the researcher will produce a final version of the questionnaire, ready for use in the real investigation.

9. *Selecting a sample of the population*

Let us suppose that the town has a population of 5,000, all of which are customers of the water company. Half of the population can see/smell the treatment works. The other half of the population also have a 'stake' in the redevelopment of the treatment works due to their exposure to potential effects such as tanker movements, traffic disruption, reduced tourist numbers, negative impact on the national park, lowered property values, etc. Thus the water company may decide that the survey population should be that of the entire town.

Ideally, all 5,000 residents would be questioned - but this would be too expensive and time consuming. Instead a sample is needed that, as far as possible, represents the entire population.

The best way of ensuring a representative sample is to use completely random sampling (also known as probability sampling). In this case the names of the residents/customers would be numbered 1 to 5,000. A random number table could then be used to select, say, 500 names for the survey sample.

Completely random sampling may be impractical for larger populations. Instead the technique of cluster sampling can be used. This allows the addresses of survey respondents/interviewees to be 'clustered' geographically - reducing costs and travel expenses (if face-to-face interviewing is used). In theory, it gives a sample as representative of the population as that achieved by completely random sampling.

It is possible that, by chance, a random sample will include a higher proportion of one group than there should be for the sample to be truly representative. E.g. if the population of 5,000 consists of 2,600 women and 2,400 men (52/48 percent) the sample of 500 might, by chance, consist of 310 women and 190 men (62/38 percent). Stratified sampling can be used to avoid this possibility. In this instance the population would be divided into men and women and a 10% sample taken from each group. This would guarantee a sample of 260 women and 240 men – the same ratio as in the population.

Quota sampling does not meet the exacting standards of random sampling but is faster and cheaper. The researcher decides how many people of each category should be included in the sample and goes out looking for them until the quota is filled. For example, the researcher may target 500 local residents who use the supermarket (in order to discover the effect of the nearby treatment works on their shopping experience) and aim for roughly equal numbers of men and women. The sample isn't truly random, as the researcher can target anybody they like within these guidelines. Despite the potential for bias, results of quota samples are often closely comparable with those from other kinds of representative samples.

10. Selecting and training interviewers (if necessary)

If interviewers are required they must be selected carefully, as their sex, class, race, or accent may all affect the answers given.

Interviewers should also be trained to avoid the dangers of 'interviewer effect'. A careful balance must be struck between establishing the kind of relationship with respondents that will encourage them to be frank and truthful, and avoiding becoming too friendly so that they try hard to please. This preferred approach might be described as 'friendly but restrained'. Interviewers are usually encouraged to be as non-directive as possible although they may sometimes pretend not to understand what is being said, in order to prompt the respondent into saying more (McNeill, p39).

11. Collecting the data

We will now consider the merits of data collection by postal questionnaire and standardised interview (face-to-face and telephone). Each method has advantages and disadvantages - neither is 'right' or 'wrong'. The choice of method will depend on its appropriateness and the resources available to the researcher.

Advantages of postal questionnaires over face-to-face, standardised interviews include:

- lower costs (data collection and processing);
- avoidance of interviewer bias;
- geographical spread of respondents is less of a problem;
- large numbers of respondents can be surveyed in a relatively short period;
- respondents can answer in their own time and can look up information if necessary;
- respondents can see the context of a series of questions.

Face-to-face standardised interviews do, however, have the following advantages over postal questionnaires:

- they generally have higher response rates, and therefore reduced biases;
- they are more suitable for respondents with poor literacy; for the visually handicapped; the very old or young; and those with language difficulties;

- they provide an opportunity to correct misunderstandings or to probe, or to offer explanations or help;
- they are better for asking numerous open-ended questions;
- they allow control over the order in which questions are answered and a check on incomplete responses, incomplete questionnaires or the passing on of questionnaires to others.

McNeill (1990) states that low response rates (usually around 30-40%) are the major drawback of postal questionnaires. He compares this with a response rate of 70-80% for face-to-face research.

Mangione, by way of contrast, recommends that researchers using postal questionnaires aim for a response rate of 75%. He proposes a number of ways to help achieve high response rates:

- send three (or more) reminders, at 2-week intervals, after the initial questionnaire;
- ensure the questionnaire is sent with a good covering letter;
- ensure that instructions for completing the questionnaire are clear;
- guarantee confidentiality – people are more likely to respond if they know their answers will not be attributed to them directly;
- consider using real stamps for return postage (rather than business reply envelopes) – they seem to produce small increases in return rates;
- consider using rewards (monetary or non-monetary) - even small ones can improve response rates. Rewards that are prepaid are likely to produce higher responses than promised rewards.

Lastly, telephone interviews may be used instead of postal questionnaires or face-to-face interviews. Like postal questionnaires, telephone interviews are relatively cheap and fast. They can also reduce resistance to 'sensitive' items.

The main disadvantage of telephone interviews is that of sample representativeness. Not all households or individuals are accessible by phone and this is usually linked with lower income, with being young and male, and with having moved house recently. In addition, many people choose to go ex-directory. Problems may also be experienced with respondents who are hard-of-hearing, elderly or from minority groups – but probably no more so than with conventional interviews.

Refusal rates for telephone interviews can be high at evenings and weekends but are generally similar to those of conventional interviews. In some countries there is increasing resistance to 'cold' telephone calling.

12. Processing the data and analysing the results

Before the data can be analysed, it first has to be converted (processed) into a numerical form by a classification system known as coding. This simply means the allocation of a number to each answer category.

'Closed', multiple-choice, questions can, and should, be pre-coded. An example is shown below.

Q. Can you see the waste water treatment works from your home?
[Please tick]
(a) Yes......... [ ] 1
(b) No...... .... [ ] 2

In this case the numbers 1 and 2 alongside the check boxes are the answer category codes. Some respondents might, however, skip this question or write "don't know" – even if there is no check box for that purpose. Numerical codes should be devised for such occurrences and used consistently throughout the survey.

'Open', free response, questions will produce data in the form of words and sentences, either written or recorded on tape. The answers to such questions cannot be pre-coded. Instead, the classification scheme ('coding frame') must be developed by studying responses from the main sample.

Once coding is complete, entry of data from each completed questionnaire into a computer can begin. Clerical errors at this stage can be reduced through the use of specialised data entry management software.

The final result of the data entry phase will be a two-way matrix of variables versus respondents i.e. the database. Data processing and statistical analysis of the database should be conducted with the aid of a 'code book' comprising of:

- the original questionnaire, together with the codes allocated to each answer category;
- the coding frames (the classification system for dealing with responses to open questions);
- the variable allocation document (this contains the text and label of every question and every answer category as well as the variable number which has been allocated to it).

Before statistical analysis is started, the data set should be 'cleaned' using a series of checks. The first such check is normally with regard to frequency distributions e.g. does the data set contain expected numbers of male and female respondents?

Secondly, range-checks should be run for each variable. E.g. if a question has answer codes 1, 2, 3, or 9, then there should be no entries of 4, 5, 6, 7, 8, 10, etc.

Lastly, internal consistency checks should be used to pick up probable inconsistencies, such as respondents answering 'no' to one question but answering the following questions as if the answer to the first question had been 'yes'.

Possible areas of statistical analysis (from "Survey Research" by Roger Sapsford):

- Tabular analysis
  - Significant association in tables
  - Three way analysis and statistical control
- Correlation
  - Multiple and partial correlation
- Regression
  - Multiple regression
  - Stepwise regression and testing effects
  - Causal modelling
- Discriminant function analysis
- Factor analysis
- Analysis of variance
  - Univariate analysis
  - Bivariate and multivariate analysis
- Loglinear analysis
  - Logistic regression
  - Loglinear modelling
  - Residuals and the analysis of change
  - Latent variables

For a more detailed discussion of suitable statistical techniques for survey research, see Barnett (1991), Clarke and Cooke (1998), Freedmann et al. (1998) and Lewis-Beck (1995).

B1.3.2 Focussed group interviews (focus groups)

Data from external stakeholders can also be collected though the use of focus groups. These generally involve "8 to 12 individuals who discuss a particular topic under the direction of a moderator who promotes interaction and assures that the discussion remains on the topic of interest" (Stewart & Shamdsani, 1990).

Focus groups are often used as a preliminary stage in a larger study. They are useful in that they provide a means of:

- obtaining general background information about a topic of interest;
- generating research hypotheses that can be submitted to further research and testing.

Consider the earlier example of the wastewater treatment works redevelopment. In this case the WSP could use focus groups to get a general feel for what people think about the existing works and what their concerns would be about any redevelopment. This data would help inform the design of any questionnaires/interview schedules used in a subsequent survey.

Even if a follow up survey is not planned, focus groups can still be used to help in "diagnosing the potential for problems with a new program, service, or product" (Stewart & Shamdasani, 1990).

A key difference between focus groups and survey research is in the type of data collected. Focus groups are almost always used to collect qualitative data. They allow individuals to respond in their own words, using their own categorisations and perceived associations. This is quite unlike the collection of quantitative data using survey research questionnaires. Such questionnaires place tight constraints on the way in which responses can be expressed.

With focus groups the researcher can obtain deeper levels of meaning, make important connections and identify subtle nuances in expression. In addition, the researcher can observe non-verbal responses such as gestures, smiles, frowns, and so forth, which may carry information that supplements the verbal response (Stewart & Shamdasani, 1990).

Focus groups have the following advantages over other types of research:

- data can be collected much more quickly and at less cost than if individuals were interviewed separately;
- researchers can interact directly with respondents (unlike survey research using questionnaires) and so can provide clarification, probe responses, etc;
- the group setting may produce data that might not have been obtained from individual interviews;
- the results of a focus group are easy to understand (unlike the complex statistical analyses of some survey research);

and the following disadvantages:

- the small number of respondents limit generalisation to a larger population;
- responses of group members are not independent – again limiting generalisation;
- focus group results may be biased by a very dominant or opinionated member;
- the 'live' and immediate nature of the interaction may lead to greater faith in the findings than is actually warranted;
- the open-ended responses from focus groups can be difficult to summarise and interpret;
- the moderator may bias results by providing cues about what types of responses are desirable.

The sequence of steps in the design and use of focus groups is as follows (Stewart & Shamdasani, 1990):

11. Problem definition / Formulation of the research question
12. Identification of sampling frame
13. Identification of moderator
14. Generation and pretesting of interview guide
15. Recruiting the sample
16. Conducting the group
17. Analysis and interpretation of data
18. Writing the report

The first step is to define the problem: what kinds of information are needed and from whom? What issue(s) will the groups actually focus on?

The second step is to identify a sampling frame (population sample) that is representative of the larger population. This is not as critical for focus groups as it is for survey research because we should not generalise too much from focus groups anyway. Nevertheless, the sampling frame should be at least a good approximation of the population of interest.

Identifying the moderator and designing the interview guide should be carried out with regard to the group to be interviewed. Moderators may be more or less suited to interviewing certain groups e.g. elderly people, male, female, different social classes, etc. Likewise, the interview guide should be tailored to suit the particular group of respondents. Unlike survey questionnaires, interview guides do not consist of a highly structured set of question and set of responses. Rather they set the agenda for a focus group discussion. As with survey research, the interview guide should be tried out (pre-tested) prior to use. This helps determine problems with the wording of questions, whether they elicit discussion and whether they are easily understood. Pre-testing should ideally be carried out with individuals who are representative of those who participating in the actual focus groups. Those involved in the pre-test should not be aware of the research purpose or have been involved in the design of the interview guide.

The next step is to recruit the focus group participants. In practice, this is often done simultaneously with the identification of the moderator and design of the interview guide. Attention must be paid to the composition of the group as this may affect both the quality and direction of the discussion. A time and place must be identified for the group and arrangements made for any special facilities or equipment needed. The focus group should be conducted somewhere that is reasonably close to participants. It is normal to inform potential participants of the general topic of the focus group and to offer them an incentive for participation. It is also normal to over-recruit for focus groups i.e. to invite a few more participants than the number needed. This compensates for participants who, for various reasons, fail to show up. Finally, it is good practice to remind participants about the focus group, by letter or telephone, a few days before it is due to take place.

Now the focus group can actually be conducted. The initial task of the moderator is to establish an environment in which participants can express themselves freely. Later, the moderator steers the groups through the questions on the interview guide and facilitates discussion among participants. As the interview proceeds, the moderator must decide on the appropriate level of directiveness, structure and intimacy. The discussion is often audio or video taped and observers may also be present. The use of taping/observers should be explained to participants before the start of the interview. When complete, the moderator should debrief participants about the purposes of the focus group.

The amount of data analysis carried out will depend on the research purpose and the costs and benefits of such analysis. Consider again the redevelopment of a wastewater treatment works, for which there are a number of options. A series of focus groups might clearly show the preferred option to be unacceptable. In this case the results of the focus group are so obvious that detailed documentation and analysis are probably not worthwhile. However, it is more common for the interview to be transcribed and for its content to be analysed. This Content Analysis can take a number of forms – ranging from simple 'cut-and-paste' techniques (where the transcript is literally cut up, categorised and interpreted) to highly sophisticated computer assisted analyses.

B1.3.3 Individual 'depth' interviews

This type of individual interview should not be confused with the standardised interview used in survey research. Instead, the depth interview provides: "the opportunity for the researcher to probe deeply, to uncover new clues, to open up new dimensions of a problem and to secure vivid, accurate, inclusive accounts that are based on personal experience" (Walker, 1985)

Depth interviews have much in common with focus groups. Like focus groups, the depth interview is relatively unstructured. While an interview guide may be used, a questionnaire certainly wouldn't. Likewise, depth interviews are normally recorded and transcripts analysed individually. Furthermore, the depth interviewer will share some of the concerns of the focus group moderator. These include the need to obtain trust, to avoid biasing responses and to establish an appropriate level of structure.

Depth interviews have some advantages over focus groups. Firstly, more time can be spent with each individual and so their position can be explored more deeply (hence the term depth interview). Secondly, individual interviews avoid the social pressures that can occur in focus groups. Those who are reluctant to give their views in front of a group may be more comfortable in an individual interview. This makes individual interviews more suitable for topics that are potentially sensitive or embarrassing. Thirdly, it is easier to organise individual interviews than to get numerous people together for a focus group.

One disadvantage of depth interviews is that they lack the social interaction of focus groups. This interaction is desirable when we want participants to take account of other people's views. The social interaction of focus groups may also help stimulate participants to a greater degree than would be the case in an individual interview. This is useful if group members are being used creatively to generate new ideas, perspectives or insights. Another disadvantage of depth interviews is that of cost: it is clearly quicker and cheaper to interview people in a group than to do so individually.

B2 INFORMATION GENERATION TOOLS

B2.1 Modelling (simulation) of existing and alternative systems

Phase 4 in the decision support process is the collection of data by which to score or rank each option according to the chosen criteria. In order to evaluate the future performance of the system under different options, data collection on the past performance needs to be combined with modelling and assessment of the system over time. There will be an interaction between the criteria chosen for assessment and the modelling tools used to evaluate future performance of the system. The tools chosen should be appropriate to model the future values of the criteria under the range of options to be considered.

A hierarchy of different types of model is available and, often, a combination of different levels of modelling will be required. These range from detailed process modelling tools, such as Hydroworks, to higher-level systems modelling approaches, such as the Reference Sustainability System. The interaction between different levels of modelling is a key feature of the assessment process.

Detailed modelling tools are essential to capture the performance of the system under specified conditions relating to each option under consideration. For example, in the sanitary waste case study, an InfoWorks model was built to simulate the dry and wet weather flow in the combined sewer system under investigation.

However, this type of model is highly data intensive and often requires long running times, making it impractical to do more than a limited number of model runs. Thus, it may be necessary to supplement detailed modelling tools with higher-level systems models, which allow more flexible assessment of the interaction between different parts of the system under a wider range of conditions. This type of model is exemplified by the Reference Sustainability System approach.

The Reference Sustainability System (RSS) approach was developed by members of the SWARD research team from Imperial College in the course of a previous EPSRC-supported project (Leach et al., 1997; Foxon et al., 1999, 2000). This was a development of the Reference Energy System (RES) approach, which is widely used for modelling energy systems (Meier, 1984). The Reference Sustainability System is a methodology for representing the resource and material flows on which the environmental sustainability of a system depends, in order to enable systematic assessment of the potential of technologies and resource management strategies to enhance the sustainability of the system. An RSS systems model aims to represent the key activities over the life cycle of a resource flow, from source, supply, distribution to end-use and waste disposal. Constraints are imposed on the model to ensure that materials balance is maintained at each stage of the system (ensuring all flows are accounted for), that end-use demands are met, and the technical capacity of systems elements is not breached. A range of options and scenarios for technological or management changes to the system can then be modelled by changing the constraints, and solving the model for the flows which meet the newly-constrained system.

Such a systems model can either be developed as a stand-alone model, requiring the collection of large amounts of data, or as a complement to a more detailed process model. In the latter case, the systems model can take the output of the finite runs of the detailed process model as a starting point and interpolate the state of the system for values between these points. Whilst this introduces another level of approximation into the modelling, it can nevertheless be useful for understanding the system behaviour under a wider range of conditions. For example, the systems model can be used to assess the results of taking combinations of the basic options, in order to estimate the optimum combination according to specified criteria. This estimated optimum combination can then be tested again using the detailed process model. This type of modelling was applied in the sanitary waste case study, as described in Appendix D4.

B2.2 Economic Analysis Tools

Water Service Providers (WSPs), whether publicly or privately owned, will seek to obtain 'value for money' from their proposed or potential investments. Consequently, alternative investment options will almost certainly be subjected to some type of economic analysis.

Economic analysis tools are important because the information they generate will often have a significant bearing on the final decision. Three important tools are:

- cost-benefit analysis;
- social cost-benefit analysis;
- cost-effectiveness analysis.

These are described below.

B2.2.1 Cost-Benefit Analysis and Social Cost-Benefit Analysis

Cost-Benefit Analysis (CBA) has been described as "an economic tool for determining whether or not the benefits of an investment or a policy outweigh its costs" (Wrisberg et al, 2002). Two types of CBA have been identified: economic and social. Economic CBA (also known as financial CBA or financial analysis) involves real cash flows that affect the investor. Social CBA (also known as extended CBA or economic appraisal) involves real and theoretical cash flows that affect the overall welfare of society (Best & De Valence, 1999). Social CBA (SCBA) allows factors beyond the direct interest of the investor to be incorporated into the analysis. These factors and externalities, may include the social and environmental impacts of the proposed development. The ability of CBA (social) to allow consideration of social, environmental and economic factors suggests it can improve the inclusion of sustainability within decision making.

In CBA the costs and benefits of alternative options are identified and quantified in monetary units. Those occurring in the future may be discounted for the purpose of "equating the value of money in different time periods" (Best & De Valence, 1999). The discounting mechanism allows the present value (PV) of any future cost or benefit to be calculated. Subtracting the PV of all costs from the PV of all benefits gives the net present value (NPV). This is a commonly used economic criterion for selecting between mutually exclusive options. The option with the highest NPV should be preferred, provided it is positive. Other economic evaluation criteria based on the same assessment of costs and benefits include benefit-cost ratio (BCR); internal rate of return (IRR); equivalent annual worth (EAW) and payback period (Dandy & Warner, 1989).

According to Best & De Valence (1999), CBA is the most popular technique for evaluating public projects and the primary technique used in the field of environmental economics. Pearce (1983) explains the appeal of CBA to those with a real-world task to attend to, such as deciding on the priorities within any sub-budget of government expenditure. To such people, it seemed that CBA offered the only reasoned technique for aiding the evaluative process, albeit subject to many caveats. CBA also had a fundamental attraction of reducing a complex problem to something less complex and more manageable.

CBA has, however, attracted a number of criticisms. Perhaps the most common objection to CBA is the requirement to express all benefits and costs in monetary terms. Self (1970), for example, has described the valuation of non-market benefits and costs as "a confidence trick – to exploit the ordinary man's respect for the yardstick of money in what are actually non-monetary situations" while Schumacher (1973) states that all CBA can do is "lead to self-deception or the deception of others" (Pearce, 1981).

Another criticism is that CBA takes no account of distributional effects or equity issues. No concern is given to who gains and loses from a project: as long as the total benefits outweigh the total costs. Likewise, there are questions over intergenerational equity. Consider the situation where a species is under threat. In this context Side (1998) argues that 'it may not be reasonable to suggest individuals of the present generation could be compensated for (or even accept on their behalf) a loss of environmental inheritance for all future generations, any more than they could pay now for this loss for all future generations.'

It has also been suggested that CBA may be manipulated in order to produce a pre-determined result and may be less useful where political decisions dominate (Best & De Valence, 1999) – although the same could no doubt be said of other analysis techniques.

Despite these problems, it is hard not to agree with Pearce & Nash (1983) when they say that: 'the discipline of CBA (or a similar formal technique) at least forces the process of evaluation to list all gains and losses and to weigh up their relative values. This may seem a small virtue. But in a world where decisions are made more often than not on irrational assessments, it could remain the single most important attribute of any calculus designed to assist the decision-making process.'

CBA Methodology
A typical CBA framework is shown (Fig. B1). Its application to Social CBA will be described although it is equally valid for economic CBA. Stages 1, 2, 3, and 6 are fairly obvious and require no further explanation. The other stages merit a more detailed description.

1. Identify total project costs and benefits

After careful consideration of the project scope, objectives and constraints a number of feasible options will have been generated. The likely costs and benefits can now be identified. It is not necessary to monetise them at this stage. Project costs are those incurred over its expected life i.e. construction, operation and disposal costs. These costs may be direct (e.g. capital, maintenance, operation, selling) or indirect (e.g. theoretical costs of negative impact on society or the environment).

Likewise, the benefits produced by a project over its expected life are considered. Again, these may be direct (e.g. generation of additional revenue, improved customer service) or indirect (e.g. improvements to the environment or social welfare).

2. Quantify total project costs and benefits

The project costs and benefits should now be quantified in monetary terms. For some costs and benefits this will be relatively straightforward. A WSP should, for example, be reasonably certain of the direct costs associated with replacing buried infrastructure or improvements to above ground assets. Similarly, the WSP should be able to estimate the financial benefits of these activities, in terms of reduced repair and maintenance costs, customer contact costs, etc.

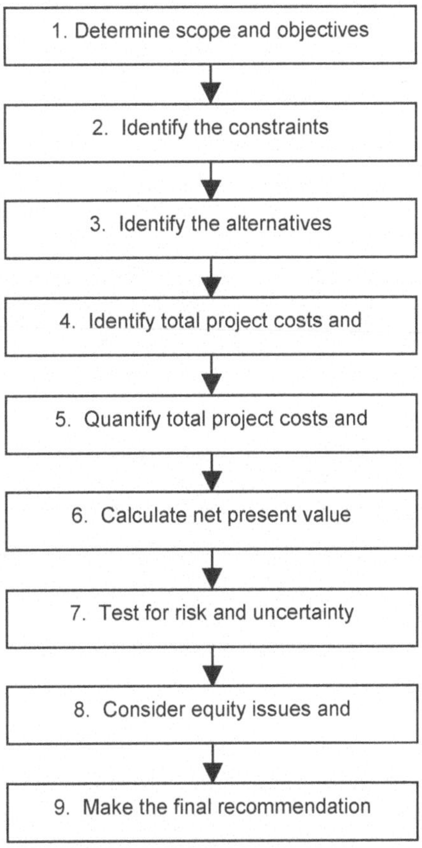

Figure B1: Cost-benefit analysis framework
(Adapted from Best & De Valence, 1999)

The monetisation of externalities, such as indirect social and environmental impacts, is more difficult. Merret (1984) has described two approaches to the monetization of environmental 'effects beyond price':

- variant costing;
- pseudo-market pricing.

Table B1 puts all of the costing techniques into a framework and shows how they are related, and Merret's explanation of these approaches is then given below.

Table B1: Relationships between costing techniques

	Observed behaviour	Hypothetic behaviour
Direct estimation	*Direct observed* Competitive market prices Simulated markets	*Direct hypothetical* Bidding games Willingness to pay questions
Indirect estimation	*Indirect observed* Travel costing Hedonic pricing	*Indirect hypothetical* Contingent valuation methods

Variant costing is largely concerned with estimating the cost of measures for the mitigation of negative environmental effects. The starting point for this technique is an Environmental Impact Assessment (EIA), on its basis it may be possible to specify project variations by which negative environmental effects can be mitigated.

Pseudo-market pricing recognises that no formal markets exist in which a project's environmental impacts are priced. Instead, indirect methods are used to estimate what people would be willing to pay, if such markets did exist. Three such methods are contingent valuation, the travel cost method and hedonic pricing.

In contingent valuation interview surveys are conducted on an appropriate sample of the population affected or likely to be affected by a project, plan or policy. The aim is to discover the maximum amount a person would be willing to pay to enjoy an environmental benefit or to avoid a dis-benefit. An anchor price is set and, providing the respondent is willing to pay more than this, the price successively raised until the maximum is arrived at. Likewise, the anchor price could be the amount the respondent would be willing to accept as compensation for the loss of an environmental benefit. These prices can then be multiplied by the size of the total population affected, to give the aggregate contingent value.

The travel cost method is concerned with the study of primary user visits to nature reserves and the like. One could carry out a survey of trips to a waterfall, for example. The cost of a trip is estimated based on entrance fee (if any), the cost of the journey and the value of the time spent making the journey. The aggregate cost is taken as the price of the revealed preference to make the visit. The survey also records the number of visits made by the respondent in the year. This price and quantity information can be used to estimate of the recreational value of the waterfall per year. Travel cost has been used a lot in transportation studies to justify improvements.

The hedonic pricing method is similar to the travel cost method in that it relates to the availability of water resources to local residents. Data is collected on house prices in a given area and the value of variables that might explain house price differentials. The data are usually cross-sectional, such as catchment with catchment, rather than time series. Multiple regression analysis is then used to estimate the price people are willing to pay, as part of the total price of the house, for each unit enjoyed of each of these independent variables. Where, for example, access to a local lake is identified as one such variable, the differential house price effect is calculable. Hedonic pricing is a form of multivariate analysis that relies on the sufficient availability of good reliable data in order to carry out the analysis. One of its drawbacks is that unless all of the significant variables have been identified you may get the wrong answer.

Merret is, however, largely critical of these techniques. While he regards variant costing as a "valuable weapon", he suggests that hedonic pricing and the travel cost method are a "waste of time and resources". He also argues that contingent valuation is based on a "category mistake" and should not be pursued. Instead he proposes a 'dual analysis' approach where CBA is combined with Environmental Impact Assessment.

In practice, it will probably be very difficult – even impossible – to obtain a credible monetary valuation of all externalities. Let us suppose that an attempt is made to identify and monetise every last impact of a project in terms of the economy, society and the environment. The monetisation process would surely become increasingly complicated, time consuming and less transparent, without necessarily gaining in credibility. Could we ever be absolutely certain that every impact had been considered?

It should be remembered that monetisation can only provide an estimate of the value of externalities, it can never be correct. Increasingly large amounts of effort spent on monetisation may provide diminishing returns in terms of improving this 'correctness'.

It may be more realistic to admit that some externalities defy monetary translation. It has been suggested that these unpriced impacts (or intangibles) are still an important ingredient in the final decision and should be included descriptively in a CBA report. Decision makers can then determine the weight to be given to these issues in their policy determination. (Best & De Valence, 1999)

3. Test for risk and uncertainty

Forecasts of costs and benefits are based on data which may itself be partially estimated and/or incomplete. This uncertainty can undermine the reliability of a CBA outcome. Sensitivity analysis is often used to test the robustness of results under different scenarios (Best & De Valence, 1999). Best and worst case assumptions for key variables (e.g. project life span, input/output quality, operation costs) are applied and the effect on the ranking of alternative options observed. If the ranking changes little (insensitive to changes) then there is greater confidence in the results of the analysis. If the ranking changes significantly (sensitive to changes) then the project has a higher risk level.

4. Consider equity issues and intangibles

The best option, using SCBA, is that with the greatest positive net benefit to society. This overlooks the fact that, within society, there are winners and losers from a project. Those projects that are likely to involve an uneven distribution of benefits should be the subject of a separate study designed to expose such effects. (Best & De Valence, 1999) These equity issues can then be considered alongside the SCBA results. Any intangibles arising from the project should also be considered at this stage.

5. Make the final recommendation

At this point, the decision-makers will consider each option in terms of its economic costs and benefits, equity issues and/or intangible impacts. Ultimately they 'must use their human wisdom to make a final judgement on whether or not a project should be financed' (Merret, 1997).

CBA Discount Rates

In CBA, discounting deals with the problem of time preference and the relative valuation of the identified costs and benefits. It is suggested that the current discount rate of 6% used in the UK, is unsuitable for long-term projects. A high discount rate may count against longer-term projects with relatively low recurring cost in favour of projects that have high recurring costs but lower capital costs. This may lead to the adoption of more unsustainable options, and impose excessive costs on future generations. Some environmentalists have even reasoned that discounting future costs and benefits may discriminate against future generations, and consequently demand a zero discount rate.

The social discount rate (SDR) is the rate used in CBA to discount the future stream of benefits and costs of the project. The concept involves attaching coefficients to the benefits and costs, and deals with intergenerational equity. The SDR, which focuses on the differential timing of effects, is used to discount future values to calculate the net present value of the stream of benefits and costs of the project. The social time preference rate (STPR) can be used as the SDR. The STPR consists of pure time preference that discounts future well-being. Individuals living today make savings decisions concerning how they wish to allocate their lifetime resources, now and in the future. The issue is then, to what extent should the STPR be based on these individual time preference rates, when the preferences of the future generations as well as those currently living must be taken into account?

There are two propositions: an individualistic or an authoritarian STPR:

- Individualistic: The preferences of the existing generation are given priority, but these preferences depend on the consumption of future generations. Individual savings may not be optimal in the presence of an externality. The problem with this approach is that no one has tried to produce (with the externality argument) an actual estimation of the STPR.
- Authoritarian: The existing generation is assumed not to be able to account for the future one, therefore, government needs to intercede and replace individual time preferences with a distinct social perspective that explicitly includes the preferences of future generations. Society should therefore be responsible for future generations as well as the current ones. Two aspects require consideration here:
    - future generations will be richer (consume more) because of economic growth
    - the additional income going to the future generation should be valued less than the additional income going to the current one as there exists a diminishing marginal social value of increases in consumption (Brent, 1998).

Economists are divided on the appropriate value of the STPR and its role in discounting. Fairly low values (0 to 5%) are generally recommended so that governments can recognize their obligation to future generations as well as to the present. The current STPR appears to be in the region of 2 - 4 % in the UK.

B2.2.2 Cost-Effectiveness Analysis

Cost-Effectiveness Analysis (CEA) has been referred to as "a modern variation of traditional CBA" (Best & De Valence, 1999) and "a halfway house on the road to CBA" (Sugden and Williams, 1978). Likewise Social Cost-Effectiveness Analysis (SCEA) has been called SCBA's "poor yet hard-worked cousin" (Merret 1997). As a lesser variant of CBA, CEA tends to merit only a few paragraphs in CBA texts.

Several definitions of CEA have been suggested. Thompson (1980) notes that CEA has been frequently defined as "benefit-cost analysis without monetary valuation of program outputs". A less succinct, but supposedly more appropriate, definition is preferred by Thompson: "Cost-effectiveness analysis evaluates a decision alternative (1) by making all effects commensurate in terms either of money or of one unvalued output unit and (2) by comparing these dimensions of impact."

As with CBA, CEA can consider environmental, social and economic factors and so may aid sustainability inclusion in decision making. Unlike CBA, CEA does not require the monetisation of project effects and so addresses one of CBA's major difficulties.

CEA is used where the outputs of alternative projects are equivalent. As the level of benefit is common, it is only necessary to calculate the costs. The project with the lowest discounted cost for the same or equivalent benefit is said to be the most cost-effective and should be preferred. Conversely, if the budget is fixed, projects are assessed in terms of comparative (non-monetary) outputs.

Of course, decision situations where alternative projects have varying costs (monetary inputs) and exactly equivalent (non-monetary) benefits are unlikely. It is more realistic for alternative projects to have different effects – good and bad – even though they fulfil the same basic function. Furthermore, some of these effects may be easily quantifiable in monetary terms (e.g. reduced power costs after an equipment refurbishment project). Thus, as Schofield (1987) says, decision-makers will be left to "weigh the comparative intangible benefits against the measured comparative costs". Although definitive solutions are not presented, Schofield suggests that "a considerable amount of useful comparative information is nonetheless made available".

It should also be borne in mind that CEA assumes all project alternatives are actually worth undertaking in the first place.

B2.3 Environmental Analysis Tools

Numerous analytical tools exist that help provide information about the environmental consequences of a decision.

The European Network on Chain Analysis for Environmental Decision Support (CHAINET) project has described, compared and assessed ten of these tools (Wrisberg et al, 2000):

1. Life Cycle Assessment (LCA)
2. Material Input per Unit of Service
3. Environmental Risk Assessment (ERA)
4. Material Flow Accounting (MFA)
5. Cumulative Energy Requirements Analysis (CERA)
6. Environmental Input-Output Analysis (env.IOA)
7. Analytical Tools for Eco-design
8. Life Cycle Costing (LCC)
9. Total Cost Accounting (TCA)
10. Cost-Benefit Analysis (CBA)

Tools 1-7 are used to generate information about the environmental consequences of a decision in physical terms. Most of this information relates to the SWARD environmental criteria of resource utilisation and environmental impact and their associated indicators. Thus, the CHAINET tools can help WSPs estimate the 'environmental sustainability' of alternative asset management options.

Tools 8-10 deal with environmental consequences in monetary terms and so could also be called economic analysis tools. Life cycle costing and total cost accounting both relate to the SWARD economic criteria of life cycle costs. Cost-benefit analysis, which has already been described some detail, does not directly relate to the SWARD economic criteria.

Some of the main points regarding these tools, from the CHAINET description, are listed below. Cost-benefit analysis is omitted because it has previously been discussed. Guidebook users are advised to obtain a copy of the CHAINET report if they require more information.

1. ***Life Cycle Assessment/Life Cycle Analysis (LCA)***

- Aims to specify the environmental consequences of products or services from cradle to grave.
- Enables a study of different options to supply a given function.
- All processes connected with the function, from extraction of resources to final disposal of waste, are considered.
- All environmental issues connected with the function are specified as resulting from extractions, emissions and other physical interventions like changes in land use.
- May provide quantitative or qualitative results.

2. ***Material Input per Unit of Service (MIPS)***

- Used to analyse the basic environmental impact potential of certain activities.
- Has two components: Material Input and Service Unit
- The total Material Input (MI) of the analysed product consists of the materials used directly or indirectly during the life cycle.
- Service means the utility or function that can be obtained from a product (or infrastructure) in order to satisfy human needs and desires.
- Assumes that the lower the material input, the lower the material flow related impact potential.

3. ***Environmental Risk Assessment (ERA)***

- Risk assessments vary widely in scope and application.
- Risk assessments may examine effects on humans (health risk assessment) and ecosystems (ecological risk assessment)
- May be quantitative or qualitative.

4. ***Material Flow Accounting (MFA)***

- Aims to specify the pathways of materials in, out and through the economy of a nation, region, community, business sector, company or household.
- Flow of bulk materials (steel, wood, etc) analysed to study industrial metabolism.
- Cause-effect modelling possible by studying the flow of a single substance (or group of substances) with specific environmental effects.
- Can link material flows with environmental flows – making it possible to link pollution problems and resource requirements to economic causes.

5. ***Cumulative Energy Requirements Analysis (CERA)***

- States the entire demand valued as primary energy which arises in connection with the production, use and disposal of an economic good (product or service) or which may be attributed respectively to it in a causal relationship.
- Primary energy is the energy content of energy carriers that have not yet been subjected to any conversion
- Primary energy requirements are measured in Joules and aggregated into one number.
- Lower values of primary energy requirements can be interpreted as being associated with less environmental burden – provided the relative share of the energy carriers will not be changed towards more hazardous ones.

6. ***Environmental Input-Output Analysis (env. IOA)***

- Is an extension of the traditional Input-Output Model dating from the 1930's.
- Traditional IOA focuses on how inter-industry trading influences overall demand for labour and capital within an economy.

- Basic distinction (in traditional IOA) made between output of goods/services sold to Final Demand (households, governments, exports, investment) and the Total Output of the various sectors (final demand + intermediate demand).
- Environmental IOA requires additional conditions to enforce consistency among inter-industry production, pollution generation and pollution abatement activities.

7. ***Analytical Tools for Eco-design***

- Three types of analytical tools for eco-design:
  - Quantitative LCA-type tools (already described)
  - Matrices
  - Checklists
- Matrices
  - Relate two or more factors to each other; the vertical part often considers the different life stages and the horizontal part the interventions, extractions and emissions.
  - Data can be quantitative, semi-quantitative or qualitative.
  - Provide a fast and easy method of showing where improvement can be achieved without performing a full LCA study of a product.
- Checklists
  - Series of questions or points of attention.
  - Used to ensure that certain aspects of design and environment are nor forgotten.
  - May also be used as an idea generation method.
  - Flow diagrams, priority lists and eco-wheels all belong to the category of checklists.

8. ***Life Cycle Costing (LCC)***

- Looks at the complete life-span of a product to calculate the entire life-cycle costs.
- Life-cycle costs include all internal and external costs.
- External costs (externalities) are not borne directly by the company.
- LCC puts a monetary value on the emissions and resource use quantified in life-cycle inventories and on the environmental and human health effects considered in impact assessments.
- Monetary values can be developed using approaches such as contingent valuation, hedonic pricing and regulators' revealed-preferences.

9. ***Total Cost Accounting (TCA)***

- Describes the long-term, comprehensive analysis of the full range of internal costs and savings resulting from pollution prevention and other environmental projects undertaken by a firm.
- TCA is a dynamic subset of LCC and is subject to change as boundaries between internal and external costs shift (due to changing regulation and company policy).
- TCA is designed to overcome the shortcomings of conventional cost analysis i.e. failure to capture the full range of benefits from environmental projects.
- Application of TCA has four key elements:
  - comprehensive costs and savings inventory;
  - more precise cost allocation;
  - use of time horizons long enough to capture long-term costs and savings;
  - use of profitability indicators which account for the time value of money.

B2.4 Social Analysis Tools

There are a range of methodologies and tools that can help with the incorporation of social data into the decision-making process. These include basic data collection techniques such as households surveys, questionnaires, focus groups and workshops, through to the more structured analytical frameworks such as Social Impact Assessment (SIA) and Social Capital

Assessment Tool (SOCAT). These frameworks can help to assess the likely impact of policy, or other decisions, on stakeholder groups, and also examine the effect that social networks and behaviour may exert on a particular decision outcome. In recent years computational approaches that use Artificial Intelligence (AI) to study social organizations and institutions have become increasingly common. A number of modelling approaches can be used to depict social networks, with the most common contemporary approach being the use of agent-based modelling (ABM). Some of the various tools are described below in more detail.

Social Impact Assessment (SIA)

Social impact assessment (SIA) refers to an analytical method that uses a range of open-ended, semi-structured, and closed data collection techniques to disaggregate the impact of policy changes on different stakeholder groups. It examines both the impact on people, the impacts of stakeholders on the reform, and how people respond to the opportunities created by the policy actions. Qualitative SIA is conducted when adequate data for economic analysis are available and permit disaggregation of reform impacts on different stakeholder groups. Under those circumstances, SIA relies on open-ended and semi-structured data collection techniques to generate qualitative results that feed into and complement economic analysis. SIA using mixed methods is conducted when quantitative data are missing or do not permit disaggregation by meaningful social categories. SIA then includes purposive surveys tailored to the reform that can be fielded with lesser time/resource intensity than household surveys.

SIA can be used to measure direct impacts of policy (*policy impact analysis*) and can also be used to examine how people respond to policy change (*opportunity analysis*). Policy impact analysis is used to assess how the cost and benefits of reforms are distributed among different stakeholders and over time. Opportunity analysis uses qualitative techniques to examine the assets and capabilities needed for a stakeholder group to gain from reform, and informs behavioural analysis; it also suggests policy measures or complementary programs that might enable excluded groups to strengthen key assets to cope with change, and take advantage of new opportunities.

Qualitative social impact assessment can be conducted using rapid field assessment, key informant interviews, structured interviews, questionnaires, semi-structured interviews, or focus groups. This could be supplemented by a household survey or where one is lacking (or insufficiently detailed), by new surveys of households in specific regions or social groups, depending on the reform. For opportunity analysis, the SIA dataset needs to be supplemented by an institutional assessment and key informant interviews in order to explore coping strategies, preferences, access to resources and information about opportunities created by the reform. Qualitative SIA by itself provides an insufficient understanding of poverty and social impacts. Best results are achieved with mixed methods integrating qualitative analyses with national survey data, supplemented as needed by purposive, reform-specific surveys. SIA normally relies on triangulation of results from multiple data collection techniques and employs stakeholder consultations to validate results.

Social impact assessment has several advantages: i) it can be done quickly as either a complement to or substitute for household survey data; ii) it can be used to inform and improve quantitative analysis, e.g. key attributes identified in an opportunity analysis can be tested in an econometric demand analysis; and iii) direct communication with stakeholders can improve their understanding of the policy reform. By making the interests of different stakeholders transparent, SIA offers a mechanism for understanding and negotiating potential conflicts among them. Even when conflict cannot be eliminated, recognition of divergent interests can lead to a consideration of how to contain or manage conflict or compensate losers. On the downside, distinguishing objective reality from biased perceptions may be difficult, and is best done by bringing together groups with conflicting opinions.

Social Capital Assessment Tool (SOCAT)
The Social Capital Assessment Tool (SOCAT) is a survey-based instrument that measures the opportunities and constraints faced by individuals and groups by focusing on the social assets and networks that determine their access to resources. Analysis of social networks leads to an understanding of the transaction costs associated with acquiring information (e.g. price levels, production innovation) and actions by members to overcome imperfect markets through social sanctions or mutual support. SOCAT is useful for estimating likely changes in productive behaviour at the household and community level in response to policy change and allows analysts to identify how networks and norms affect this behavioural response. It can be tailored to specific policy domains or used as a complement to give depth to other methods of data gathering and analysis.

SOCAT utilizes qualitative and quantitative instruments, including a household survey, a community questionnaire, and an organisational interview. It is not a substitute, but rather a complement, to more conventional data gathering and analysis. SOCAT is a useful tool for gaining an understanding of the underlying social networks in a society, and their likely impact on the success of a reform. As a stand alone tool, SOCAT can be data- and time-intensive. Using SOCAT to determine the effects of a specific policy reform requires that the questionnaire be adapted to the policy issue and the local context. This may require prior fieldwork, using open-ended discussions with individuals and groups, to identify the relevant dimensions of social capital for a given policy objective.

Agent Based Modelling
Agent-based modelling (ABM) is an emerging technique that examines the interaction of a number of (different) objects. It is suited to the study of complex nonlinear systems; and can thus be applied to the analysis of flows and networks. Agent-based simulation methods are proving increasingly attractive to engineers and scientists seeking to take socioeconomic factors into account in the analysis of such issues.

Agent-based social simulation (ABSS) is a form of modelling of complex adaptive systems, in which the agents within such a system are represented explicitly and individually within the model. The model agents will typically represent human individuals, but may also represent companies, other institutions or states. ABSS models are generally created using an object-oriented programming language, allowing each agent to be implemented as an object. Complex agents capable of modelling essential characteristics of human beings in connection with their decision-making and behaviour control, the interactions between individuals and the interactions of individuals with their environment are of central interest (Schmidt, 2000; Urban and Schmidt, 2001). As ABSS, and other micro simulation techniques, explicitly attempts to model specific behaviours of specific individuals, it may be contrasted to macro simulation techniques that are typically based on mathematical models where the characteristics of a population are averaged together and the model attempts to simulate changes in these averaged characteristics for the whole population. Thus, in macro simulations, the set of individuals is viewed as a structure that can be characterized by a number of variables, whereas in micro simulations the structure is viewed as emergent from the interactions between the individuals.

Such modelling techniques have been applied to water management, allowing different scenarios of actor's behaviour to be simulated and for the results to be compared with observed phenomena (Tillman et al., 1999, 2001; Downing et al., 2000). The EU funded FIRMA project (freshwater integrated resource management with agents) aimed to improve water resource planning through the development and application of ABM. The project partly focused on the social processes of water management, which has lead to the consideration of a wider range of aspects of the environment in decision-making. The models aim to improve on existing integrated assessments by explicitly representing customers, suppliers, and government and their interaction at various levels of aggregation, providing new tools for policy makers concerned with issues such as waste water, water scarcity and integrated catchment planning. ABM and associated modelling techniques are not without their critics, however. Reservations include the lack of reality, with models produced being "inspired by" rather than based on data, and the way in which a set of data is used for a simulation may be something of a "dark art" (Chattoe and Gilbert, 1997).

REFERENCES

Barnett, V. (1991), Sample Survey: Principles And Methods, Arnold, London, ISBN 0 3405 4553 4.

Best, R. and De Valence, G. (1999), Building in value, Arnold, London, ISBN 0 340 74160 0.

Brent, R.J. (1998), Cost-benefit analysis for developing countries. Edward Elgar Publishing Limited, ISBN 1 85898 567 6.

Dandy, G. C. and Warner, R. F. (1989), Planning and design of engineering systems, London: Unwin Hyman Ltd, ISBNs 0-04-620017-7 & 0-04-620018-5

Chattoe, E. and Gilbert, N. (1997), A Simulation of Adaptation Mechanisms in Budgetary Decision Making, in Conte, R. et al. (eds.) Simulating Social Phenomena, Springer-Verlag, Berlin, ISBN 3 5406 33294.

Clarke, G.M. and Cooke, D. (1998), A Basic Course in Statistics, 4th Edition, Arnold, London, ISBN 0 3407 1995 8.

Downing, T.E., Moss, S. and Pahl-Wostl, C. (2000), Understanding Climate policy using participatory Agent-based Social Simulation. In: Moss, Scott and Paul Davidsson (Volume Eds.) (2001), Multi-agent-based Simulation: Second international Workshop, MABS 2000. Boston, MA, USA, July.

East of Scotland Water (1999/2000), Annual Report and Accounts.

Foxon, T. J., Butler, D., Dawes, J., Hutchinson, D., Leach, M., Pearson, P. and Rose, D. (2000), An assessment of water demand management options from a systems approach, Journal of Chartered Institute of Water and Environmental Management 14, pp. 171-178

Foxon, T. J., Leach, M., Butler, D., Dawes, J., Hutchinson, D., Pearson, P. and Rose, D. (1999), Useful indicators of urban sustainability: some methodological questions, Local Environment 4(2), pp. 137-149.

Freedman, D. Pisani R. and Purves R. (1998), Statistics. 3rd Edition, W W Norton & Co, ISBN 0 3939 7083 3.

Leach, M., Bauen, A. and Lucas, N. (1997), A systems approach to material flows in sustainable cities: Case study of paper, Journal of Environmental Planning and Management, 40(6), pp. 705-723.

Lewis-Beck, M.S. (1995), Data Analysis: An Introduction, Sage, London, ISBN 0 8039 5772 6.

Mangione, T.W. (1995), Mail Surveys: Improving the Quality, Sage, London, ISBNs 0 8039 4662 7 and 0 8039 4663 5.

McNeill, P. (1990), Research Methods, Routledge, London, ISBN 0 415 04126 0.

Meier, P. (1984), Energy Systems Analysis for Developing Countries, Lecture Notes in Economics and Statistics, 222, Springer-Verlag, Berlin.

Merret, S. (1997), Introduction to the economics of water resources – an international perspective, UCL Press, London, ISBNs 1 85728 636 7 HB & 1 85728 637 5.

Oppenheim, A.N. (1992), Questionnaire Design, Interviewing and Attitude Measurement, Pinter, London, ISBN 1 85567 043 7 HB and 1 85567 044 5 PB.

Pearce, D. W. (1983), Cost-benefit analysis, Macmillan, London, ISBN 0-333-35281-5

Pearce, D. W and Nash, C.A. (1981), The social appraisal of projects - a text in cost-benefit analysis, Macmillan, London, ISBN 0 333 19303 2 HB and 0 333 19304 0 PB.

Schmidt, B. (2000), The Modelling of Human Behaviour. SCS Publications, ISBN 1 565 55211 3.

Schofield, J. A. (1987), Cost-benefit analysis in urban & regional planning, Unwin Hyman, London, ISBN 0 04 338145 6.

Schumacher, E.F. (1973), Small is beautiful. Blond, London, reprinted September 1989, Perennial, ISBN 0 06091 630 3.

Self, P. (1970), Nonsense on stilts: the futility of Roskill, New Society, 2 July.

Side, J. (1998), Values in the Coastal Zone – an Alternative Approach, pp45-57 in Standing Conference on the Problems Associated with the Coastline (SCOPAC, Proceedings of the Seminar on Intangible Benefits and Coastal Defence. 4 November 1998, Chichester.

Stewart, D. W and Shamdasani, P. N. (1990), Focus groups: theory and practice, Sage, London, ISBN 0 8039 3389 4 HB and 0 8039 3390 8 PB.

Sudgen, R. and Williams, A. (1978), The principles of practical cost-benefit analysis, Oxford University Press, Oxford, ISBN 0 19 877040 5 HB and 0 19 877041 3 PB.

Thompson, M. S. (1980), Benefit-cost analysis for program evaluation, Sage, London, ISBN 0 8039 1483 0 and 0 8039 1484 9 PB.

Tillman, D., Larsen, T.A., Pahl-Wostl, C. and Gujer, W. (1999), Modelling the actors in water supply systems. Water Science and Technology 39(4), pp. 203–211.

Tillman, T. Larsen, T.A., Pahl-Wostl, C and Gujer, W. (2001), Interaction analysis of stakeholders in water supply systems, Water Science & Technology 43(5), pp. 319–326.

Urban, C. and Schmidt, B. (2001), PECS - Agent-Based Modelling of Human Behaviour, in: Emotional and Intelligent II - The Tangled Knot of Social Cognition, AAAI Fall Symposium, North Falmouth, MA. AAAI Press.

Walker, R. (1985), Applied qualitative research, Gower Publishing, Aldershot, ISBN 0 566 00897 I HB and 0 566 00898 X PB.

Woodhouse, J. (1999), Cost/Risk Optimisation, Operational Reliability Document, The Woodhouse Partnership Limited.

World Bank (2002), Guide to Poverty and Social Impact Analysis, April.

Wrisberg, N., Udo de Haes, H. A., Triebswetter, U., Eder, P. and Clift, R. (Eds.) (2002), Analytical Tools for Environmental Design and Management in a Systems Perspective: The Combined Use of Analytical Tools. Kluwer Academic Press, Dordrecht, The Netherlands, ISBN 1 4020 0453 2.

Appendix C: Multi-criteria Analysis Techniques and their Application

Contents

INTRODUCTION

As referred to in Section 2.5 of this guidebook, five prominent MCDM methods have been identified, as summarised in **Table C1**.

Table C1: Comparisons of the various MCDM methods

Method	Description
AHP	The essential idea is to find the attribute trade-off weights through pair-wise comparisons of attributes and, in addition to find the value of each decision alternative for each given attribute through pair-wise comparisons of decision alternatives on that attribute.
PROMETHEE	An outranking method, based on the ELECTRE procedure, which utilises the decision maker's preferences. The two versions of PROMETHEE differ by providing either a partial or total pre-order of the alternatives.
ELECTRE	An outranking method where the predominant idea is that an alternative can be eliminated if it is dominated by other alternatives. The various versions of ELECTRE differ in their use of thresholds, weights, outranking relations and a credibility index.
MAUT	Based on the construction of a decision-maker's utility function in order to represent his preference structure.
SMART	Simple version of the MAUT procedure.

The first three methods are highly mathematical and require the support of decision analysis experts and specialist software. These techniques provide academic rigour in the analysis but the black box nature of the tools, from the perception of most stakeholders, limits their value to the wider communication of sustainable decisions. The final two methods are more easily applied and both, (the SMART method in particular) require less specialist knowledge than the previous methods. They are also either transparent, or in the case of the MAUT, opaque.

Besides the above analytical methods, decision-makers may use a less formalised, more flexible, procedural tool such as Value Management (VM). An overview of this method can be found in Section 2.5 of the guidebook. More detail, including VM evaluation methods, is given at end of this appendix. The selection of the choice of methods must give due regard to the magnitude of the decision under consideration and the extent that the outcome of the decision making process requires to be communicated to the various stakeholders.

C1 BACKGROUND TO THE CASE STUDY

To demonstrate the application of multi-criteria analysis (MCA), the main points of the principal MCDM approaches are highlighted in respect to an illustrative example. The example deals with the escape of domestic sanitary waste from the sewer system in a real catchment. A number of possible options for dealing with the escape of sanitary waste have been generated and are presented in **Table C2**.

Table C2: Options for dealing with sanitary waste

Option	Measures	Primary objective
1	Install 6 mm screens on overflows at WwTP (6 mm already installed on CSO)	To screen CSO/overflows to 6 mm or equivalent to comply with minimum aesthetic pollution requirements for discharges
2	'Think Before You Flush' (TBYF) campaigns	To encourage a change in disposal habits from WC to bin
3	Install flow storage	To reduce CSO/overflows spills thus reducing the number of SW items discharged to the environment
4	Retrofit stormwater source control	To reduce stormwater entry to piped system, thus reducing CSO spills
5	Sewer rehabilitation	To limit infiltration to piped system, thus reducing CSO spills
6	Retrofit outlet chokes on existing WCs and introduce these to new developments	To force a change in disposal habits from WC to bin

The Criteria

The criteria fall into the following broad categories:

- Economic
- Social
- Environmental
- Technical

These are summary criteria, and for a measurable quantity need to be broken down into more detail. For the purpose of this illustrative case study, **Table C3** shows the list of primary criteria, along with their group association, but it has to be borne in mind that this is not an exhaustive list.

Table C3: Description and code of criteria used in case study examples

Criteria	Code	Description
Economic		
Life Cycle Costs	**LCC**	Covers all stages of the life cycle from resource extraction, production, end-use and end-of-life, including capital, operation and maintenance components and marginal, average and total cost categories.
Financial Risk Exposure	**FRE**	Relates to the risk of loss to the company associated with an investment.
Social		
Acceptability to Stakeholders	**AcSt**	Covers acceptability and perceived impacts on health and environment.
Public Understanding & Awareness	**PUA**	Covers awareness of implications of behaviour and consumer information.
Environmental		
Resource Utilisation	**ReUt**	Covers use of water resources, land, energy, materials and chemicals.
Other	**Othe**	Covers range of environmental factors including service provision.
Technical		
Performance	**Perf**	Includes the quality of supplied water and treated wastewater.
Reliability	**Reli**	Includes the availability and distribution of water supplied and wastewater, water usage restriction and pressure of mains water, and system failure in providing these services, including flooding from sewers.
Flexibility & Adaptability	**FlAd**	Covers the ability to make future changes to the system.

It has to be noted that these are primary criteria, and may consist of a number of secondary criteria, which in turn may consist of a measurable indicator. **Figure C1** shows the relationship between the sustainability principles, the criteria and indicators.

Figure C1: Relationship between sustainability principles, criteria and indicators

SUSTAINABILITY PRINCIPLES	**Primary Criteria**	**Secondary Criteria**	**Example Indicators**
Economic			
	Life Cycle Costs		
		Resource Extraction Costs	Marginal capital costs of resource extraction
		Production Costs	Average operational costs of production
		End-use Costs	Total maintenance costs in use
		End-of-life Costs	Total end-of-life costs of remediation
	Financial Risk Exposure		
Social			
Environmental			
Technical			

C2 ANALYTIC HIERARCHY PROCESS (AHP)

Developed by Saaty (1980) in the early 1970s, the AHP provides a logical framework to determine the benefits of each alternative. The essential idea of the AHP procedure is to perform pairwise comparisons (the relative importance of one criteria, of both qualitative and quantitative types, over another) to find the attribute trade-off weights, and in addition, to find the value of each decision alternative for each given attribute (also through pairwise comparisons on that attribute). The approach used by the AHP is to have the decision-maker compare, using informed judgements, every pair of attributes and determine the relative preference of one over the other using some standard numeric scale.

The information for the AHP is arranged in a hierarchical tree (of criteria and alternatives) and then synthesised to determine the relative rankings of alternatives. The initial values of the pair-wise comparisons can be inputted into an $n \times n$ matrix, which can then be used to obtain the ranking of priorities by the use of eigenvectors and has been mathematically proven, by the author, to the best approach. Simply put, the eigenvectors can be solved by:

- a short computational way to obtain this ranking is to raise the pair-wise matrix to powers that are successfully squared each time;
- the row sums are then calculated and normalised; and
- the computer is instructed to stop when the difference between these sums in two consecutive calculations is smaller than a prescribed value.

However, there is a commonly known problem with the AHP, that in rank reversal, which is caused by the addition or deletion of alternatives. This can cause the ranking of the alternatives to change from being the least preferred to being the most preferred. There is much literature in highlighting this problem and how to deal with it, which can lead to further confusion as to which solution might be the best in dealing with this problem.

C3 ELECTRE

ELECTRE was introduced by Roy (1991) in the early 1990s to incorporate the fuzzy and vague nature of decision making and uses an outranking approach based on pair-wise comparisons. It ranks the different options in relation to the objective and criteria and is useful where the alternatives remain uncertain or where the data may be imprecise.

The nature of ELECTRE is such that it seeks a solution, which is at least good (by the ranking of preferences) as another solution for the majority of the criteria, and which is not too incompatible for the other criteria. ELECTRE incorporates values of weights and thresholds as part of its calculation procedure, both of which are imperatively important to be (correctly) determined to achieve the optimal functionality of the procedure. The thresholds, of which there are preference, indifference, and veto are found on an arbitrary basis but various recommendations and/or general rules of the thumbs for all three threshold values can be found in the literature.

Meanwhile, the weights are used to indicate the significance of a criterion to the decision-maker (and which individual weightings can be collectively analysed to obtain a weight value for each criterion for all decision-makers). However, a note of caution has to be applied with the weights, as the weights used in ELECTRE are quite different from the weights used in other decision analytical modelling approaches. In ELECTRE, the weights are "coefficients of importance" whereas in the other decision analytical models are substitution rates and assess the relative preference among the criteria.

The main differences of the ELECTRE procedure (over other decision analytical procedures) are that it introduces the concept of outranking and its use of the threshold values as part of its algorithmic procedure. Also, it is a non-compensatory procedure in the sense that a bad score on a criterion cannot be compensated by a good score on another criterion i.e. there is no trade-off on the other values as a result of one poor score.

In the context of the current project, the advantages of the ELECTRE method is that it has been widely applied in environmental and energy management problems and there are many studies of a similar nature. The literature, in this sense, can be used to provide guidelines and "rules of the thumb" to the benefit of the urban water systems sustainability project.

However, for non-experts, ELECTRE can be seen very much as a black-box methodology and the technical aspects are difficult to fully understand. In this context there would be a need for a decision analyst to work alongside domain experts to obtain the best results from the procedure – it is very rare to get an expert that specialises in the two vastly different domains such as decision analysis and environmental engineering. More generally, from the evidence of the literature, there is a misconception as to how the threshold and weight coefficient values should be obtained/calculated. In many case studies these are simply determined by 'trial and error' or estimated.

C3.1 ELECTRE case example

The information used in this illustrated example, using nine second-level criteria and six options, are shown in **Tables C4 and C5**:

Table C4: Category and code of case study example criteria

Criteria	Code	Group
Performance	Perf	Technical
Reliability	Reli	Technical
Flexibility and Adaptability	FlAd	Technical
Life Cycle Costs	LCC	Economic
Financial Risk Exposure	FRE	Economic
Acceptability to Stakeholders	AcSt	Social
Public Understanding and Awareness	PUA	Social
Resource Utilisation	ReUt	Environmental
Other	Othe	Environmental

Table C5: Code of case study exanple options

Options	Code
Install 6 mm screens on CSO and Overflows	In6S
'Think Before You Flush' campaigns	TBYF
Install storage	InSt
Retrofit stormwater source control	RSSC
Sewer rehabilitation	SeRe
Retrofit outlet chokes on existing WCs and introduce these to new developments	ROC

Table C6 synthesises the whole of the data necessary to construct a fuzzy relation of outclassing in MODE 1. For each criterion, one indicates:

- by type of threshold, values of the coefficients α and β;
- the direction of variation of the preferences (growing or decreasing with the evaluations);
- mode of calculation of the thresholds (direct thresholds or opposite);

Table C6: ELECTRE data entry for case study example

Criterion Code	Direction of Preferences	Mode of Definition	Indifference Coefficients	Preference Coefficients	Veto Coefficients	Criterion Weight
Perf	Decreasing	Direct	0/1	0/19	-	8
Reli	Decreasing	Direct	0/0.2	0/0.3	-	9
FlAd	Increasing	Direct	0/0.39	0/2	-	7
LCC	Decreasing	Inverse	0/10000	0/20000	-	10.5
FRE	Decreasing	Direct	0/0.2	0/2	-	3.5
AcSt	Increasing	Direct	0/1	0/10	-	4.5
PUA	Increasing	Direct	0/0.3	0/2	-	3
ReUt	Decreasing	Direct	0/99	0/100	-	3.5
Othe	Decreasing	Direct	0/50000	0/50000	-	3.5

Table C7 shows the performance of each of the criteria relating to each of the options.

Table C7: Option related performance of case study example criteria

	Criteria								
Options	**Perf**	**Reli**	**FlAd**	**LCC**	**FRE**	**AcSt**	**PUA**	**ReUt**	**Othe**
In6S	5.38	0	0.6	68000	0.3	68.8	0	260	16400
TBYF	36.86	0.75	0.6	10000	0	85.9	1	16.7	806
InSt	52.70	0.15	0	70000	1	62.5	0	1410	150000
RSSC	89.21	0.4	1	3925	0	51.6	1	216	7230
SeRe	70.30	0.15	0	39000	0	89.1	0	39.7	3130
ROC	6.54	1	0	32446	1	65.6	1	148	4590

Using ELECTRE III with distillation coefficients of α = -0.15 and β = 0.3.

C3.2 Distillations

The ranking algorithm of ELECTRE III uses the credibility matrix to build two rankings using descending and ascending distillation: descending distillation selects at first the best alternatives to end the process with the worst ones. On the contrary the ascending distillation selects first the worst alternatives to end the process with the best ones. Two complete pre-orders are therefore found on all the alternatives.

An alternative which is incomparable to a group of others will be positioned at the end of this group in the descending distillation and at the top in the ascending distillation.

The descending and ascending distillations for the illustrative example are shown in **Figure C2**.

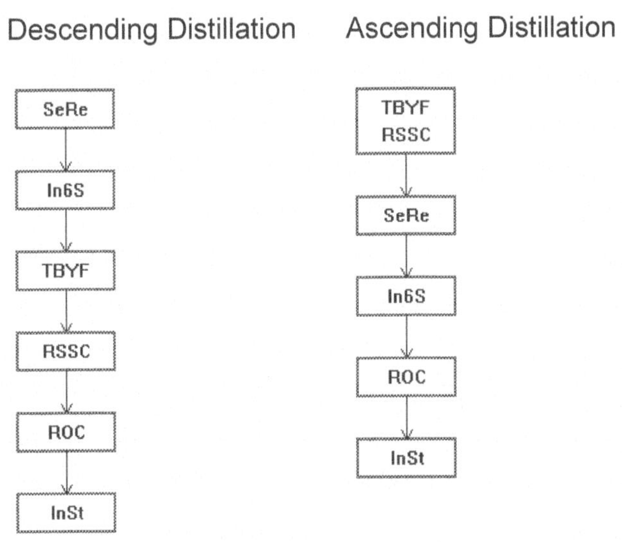

Figure C2: Ascending and descending distillations

After the two distillations, the software has built two pre-orders on all the alternatives. In order to highlight the possible incomparabilities between alternatives, the method builds the pre-order which is the intersection of the first two.

- Alternative a will be considered as better than alternative b, if in at least one of the rankings a is ranked before b and if in the other a is at least as good as b.
- The alternative a will be considered equivalent to alternative b if the two alternatives belong to the same equivalence class in the two pre-orders.
- Alternatives a and b are incomparable if a is better ranked than b in ascending distillation and b is better ranked than a in descending distillation or vice-versa.

As a conclusion, in the final preorder, the comparison between two alternatives lets 4 cases appear :

- a may be better than b,
- b may be better than a,
- a and b may be equivalent,
- a and b may be incomparable.

The ranks of the alternatives are built in the following manner: any alternative which has no better alternative will have rank 1 (even if it is incomparable to many others), the alternatives of ranks two are those whose better alternatives are only of rank 1, the alternatives of rank 3 are those whose better alternatives are only of rank 1 or 2 and so on.

This representation completely hides incomparabilities: two alternatives with the same ranks may either be equivalent or incomparable and there is no way to discriminate between these two possibilities. In the same way for alternatives with different ranks, it is not possible to know if an alternative is better than another one or if they are incomparable. These different cases are clearly distinguished in the display of the matrix of the final pre-order or in the graph.

The final pre-order ranking is shown:

Table C8: Ranks in final pre-order

Rank	Alternative
1	TBYF
	SeRe
2	In6S
	RSSC
3	ROC
4	InSt

C3.3 Median Pre-order

This ranking is another possibility for the users who do not wish to take into account incomparabilities. This ranking gives a complete pre-order (that is that any two alternatives are comparable: one being better than the other or the two being equivalent) built from the final partial pre-order in the following manner: the alternatives are ranked following the ranks in the final pre-order then the alternatives having the same rank are distinguished according to their rank difference in the two pre-orders (it indicates the relative stability between the two pre-orders).

The median pre-order ranking for the case example is shown in **Table C9**:

Table C9: Ranks in median pre-order

Rank	Alternative
1	TBYF
	SeRe
2	In6S
3	RSSC
4	ROC
5	InSt

C3.4 Concordance Matrix

The first option Concordance Matrix displays all the comprehensive concordance indices for every pair of alternatives (a, b). The comprehensive concordance index c(a, b) is the sum of the concordance indices cj(a, b) on each criterion weighted by the weights of each criterion.

cj(a, b) is calculated from the comparison of alternatives a and b on criterion j:

- if the performance of a is greater or equal to the one of b or if the performance of a is smaller to the one of b but a staying indifferent to b then cj(a, b) = 1,
- if b is weakly preferred to a: cj(a, b) is obtained with an linear interpolation and is between 0 and 1,
- if b is strictly preferred to a then cj(a, b) = 0.

Then, the comprehensive concordance index c(a, b) is calculated in the following manner:

$$c(a, b) = \Sigma\, kj^{*}cj(a, b) / \Sigma\, kj$$

The concordance matrix is shown in **Table C10**:

Table C10: Concordance matrix of case study example

	In6S	TBYF	InSt	RSSC	SeRe	ROC
In6S	1	0.62	1	0.77	0.64	0.71
TBYF	0.68	1	0.83	0.83	0.81	0.85
InSt	0.63	0.38	1	0.56	0.55	0.47
RSSC	0.59	0.70	0.68	1	0.61	0.76
SeRe	0.83	0.61	0.86	0.73	1	0.82
ROC	0.77	0.51	0.83	0.55	0.65	1

C3.5 Credibility Matrix

In the case of an ELECTRE III type project, the credibility degrees are obtained for the concordance indices taking into account the notion of discordance. The credibility degree d(a, b) is calculated by weakening the concordance index c(a, b) with a formula using discordance indices. These indices are calculated using all the discordant criteria (those for which b is preferred to a) and the values of the veto thresholds.
The credibility matrix for the example is shown in **Table C11**:

Table C11: Credibility matrix of case study example

	In6S	**TBYF**	**InSt**	**RSSC**	**SeRe**	**ROC**
In6S	1	0.62	1	0.77	0.64	0.71
TBYF	0.68	1	0.83	0.83	0.81	0.85
InSt	0.63	0.38	1	0.56	0.55	0.47
RSSC	0.59	0.70	0.68	1	0.61	0.76
SeRe	0.83	0.61	0.86	0.73	1	0.82
ROC	0.77	0.51	0.83	0.55	0.65	1

C3.6 Ranking Matrix

The matrix of the final pre-order offers a synthetic view of the results of the ranking method just as the Final Graph. These are two different representations of the same final partial pre-order. The alternatives are ranked in the order they have been initially defined. The couple **(a, b)** verifies one of the 4 following relations:

- if a is better than b, the symbol at the intersection of the row for a and the column for b is P,
- if a is equivalent to b, the symbol I,
- if a is as good as b, the symbol P-,
- if a is incomparable to b, the symbol R.

The ranking matrix for the example is shown in **Table C12**:

Table C12: Ranking matrix of case study example

	TBYF	**SeRe**	**ROC**	**In6S**	**InSt**	**RSSC**
TBYF	I	R	P	R	P^-	P
SeRe	R	I	P	P	R	P
ROC	P^-	P^-	I	P^-	P^-	P^-
In6S	R	P^-	P	I	R	P
InSt	P	R	P	R	I	P
RSSC	P^-	P^-	P	P^-	P^-	I

C3.7 Final Graph

In order to lighten the display, transitivity edges are omitted. If there are consecutive edges from alternative **a** to alternative **b**, it is possible to say that **a** is better than **b** in the pre-order obtained. Several equivalent alternatives are displayed in the same box. Two alternatives that are not related by an edge or a succession of consecutive edges are incomparable.
The final graph is presented in **Figure C3:**

Figure C3: Final rankings of case study example

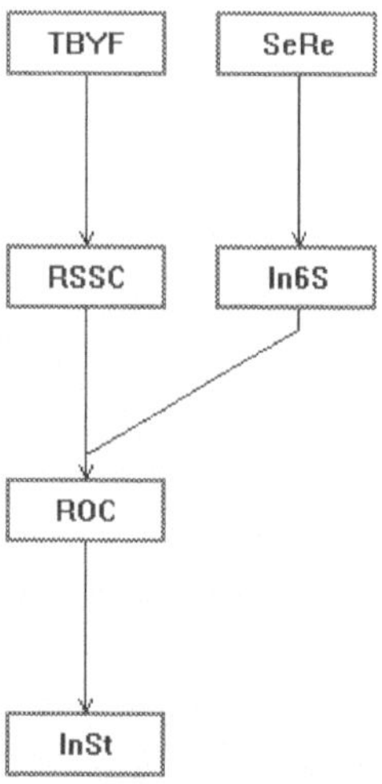

This illustrates that, given the performance matrix shown earlier and using ELECTRE III, the 'Think Before You Flush' campaign and the Sewer Rehabilitation options were deemed to be the most feasible options in the sustainability of water/wastewater systems. Due to the nature of the data, and the weighting of the criteria, the TBYF and SeRe options were incomparable to each other i.e. a definite decision of which of the two options was the better could not be determined.

C4 PREFERENCE RANKING ORGANISATION METHOD FOR ENRICHMENT EVALUATION (PROMETHEE)

Developed by Brans et al. (1984) in the mid 1980s, and like ELECTRE, PROMETHEE is an outranking method which is well supported by the literature. PROMETHEE starts with an evaluation matrix in the same vein as ELECTRE, with each cell in the matrix containing an evaluation of the criteria on the alternative. From the evaluation matrix, which each criteria can also be weighted, a preference function for each criteria has to be defined. Six possible shapes of preference functions are available and are shown in Figure C4.

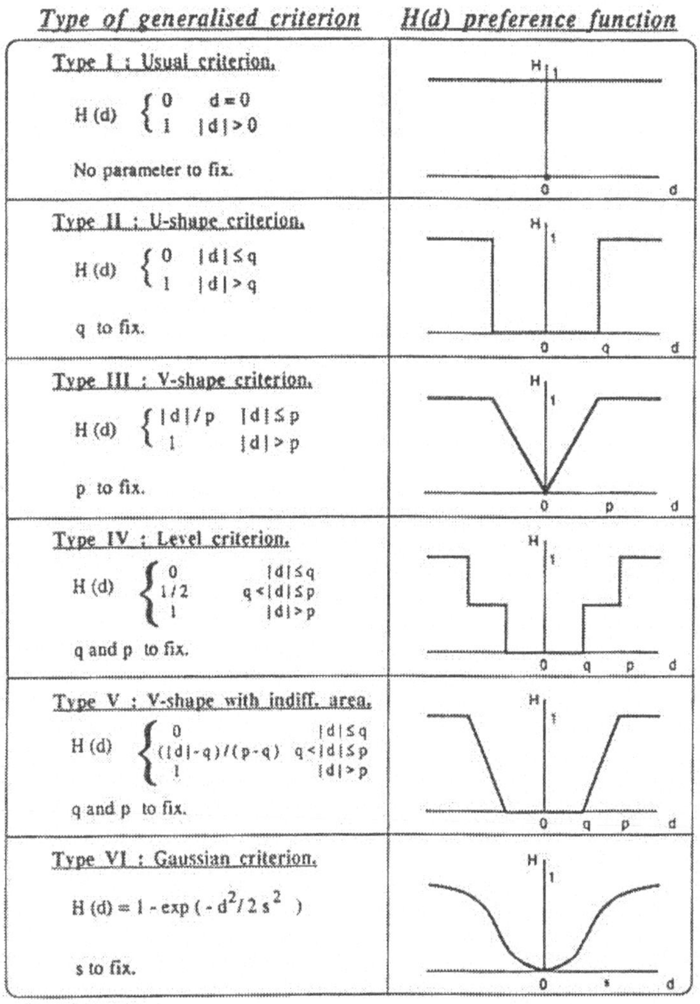

Figure C4: Overview of the six possible types of generalised criteria

PROMETHEE calculates positive and negative preference flows for each alternative. The positive flow is expressing how much an alternative is dominating the others, and the negative flow how much it is dominated by the others. The higher the positive flow and the lower the negative flow of an alternative, the better the alternative is. Based on these flows, PROMETHEE I can then obtain a partial ranking. Meanwhile, PROMETHEE II can obtain a complete ranking which is based on the balance of the two preference flows. The information from PROMETHEE II is stronger but some information gets lost in the process and is therefore more disputable. The decision-maker can use both PROMETHEE I and II in order to come to a finalised decision.

The emphasis of PROMETHEE is put on visual tools, and also automatically illustrates the sensitivity analyses by calculating the robustness of the current classification for each criterion.

C4.1 PROMETHEE Case Example

Using the same criterion and options as in the ELECTRE example, the information used in this PROMETHEE example is illustrated in **Tables C13 and 14**.

Table C13: PROMETHEE data entry for case study example

Criterion Code	Min/Max	Type	p	q	s	Criterion Weight
Perf	Min	III	19	1	1	8
Reli	Min	IV	0.3	0.2	1	9
FlAd	Max	II	2	0.39	1	7
LCC	Min	V	20000	10000	1	10.5
FRE	Min	II	2	0.2	1	3.5
AcSt	Max	III	10	1	1	4.5
PUA	Max	II	2	0.3	1	3
ReUt	Min	V	100000	99000	1	3.5
Othe	Min	II	2	50000	1	3.5

Table C14: Option related performance of case study example criteria

	Criteria								
Options	**Perf**	**Reli**	**FlAd**	**LCC**	**FRE**	**AcSt**	**PUA**	**ReUt**	**Othe**
In6S	5.38	0.0000	0.6000	68000.00	0.3000	68.8000	0.0000	260000	16400
TBYF	36.86	0.7500	0.6000	10000.00	0.0000	85.9000	1.0000	16700	806
InSt	52.70	0.1500	0.0000	70000.00	1.0000	62.5000	0.0000	1410000	150000
RSSC	89.21	0.4000	1.0000	3925.00	0.0000	51.6000	1.0000	216000	7230
SeRe	70.30	0.1500	0.0000	39000.00	0.0000	89.1000	0.0000	39700	3130
ROC	6.54	1.0000	0.0000	32446.00	1.0000	65.6000	1.0000	148000	4590

C4.2 Preference Flows

PROMETHEE I Partial Ranking

The leaving (positive) and entering (negative) preference flows are used for the PROMETHEE I partial ranking.

- The higher the leaving flow, the better the action.
- The lower the entering flow, the better the action.

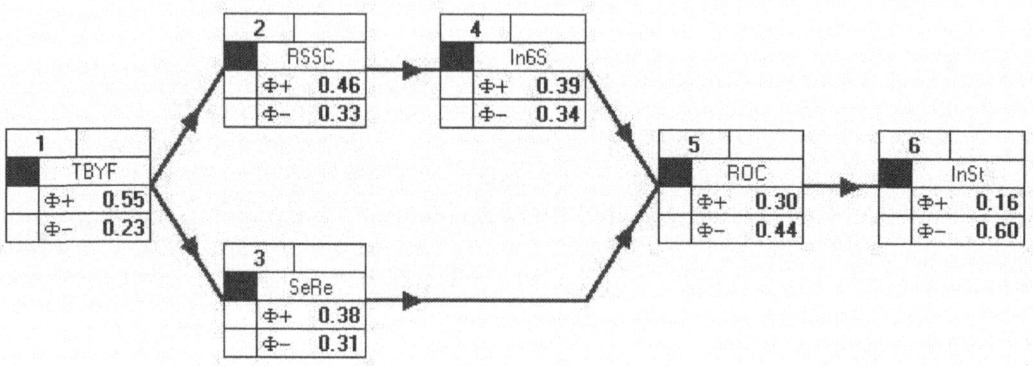

Figure C5: Graphical display of the PROMETHEE I partial ranking

The arrows indicate preferences. If two actions are not linked by one or a path of several arrows, they are incomparable in the PROMETHEE I ranking. Usually, that means that they have rather different profiles and should be considered more closely inspected by the decision-maker if he wants to select one of these.

PROMETHEE II Complete Ranking

The net flow is the balance of the leaving and entering flows. It is used for the PROMETHEE II complete ranking.

- The higher the net flow, the better the action.

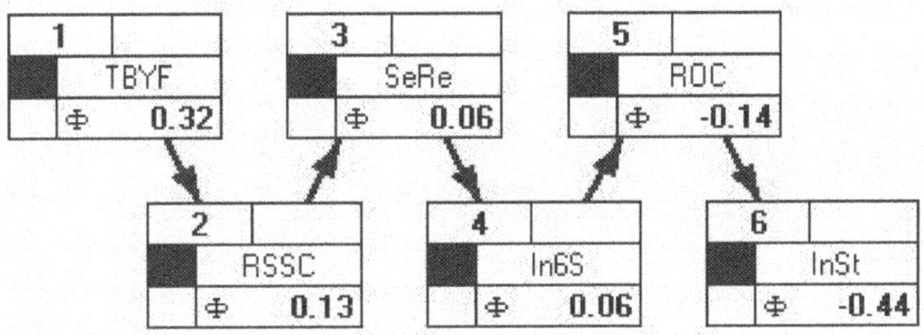

Figure C6: Graphical display of the PROMETHEE II complete ranking

The actions are graphically represented with the first ranked located to the left. The rank and actual value of the net flow are given for each action.

The values of the net flow are also graphically displayed in the lower window. In this way, the deviations between the successive values appear clearly and can be used to moderate the ranking. For instance, if two actions have net flows that are very close to each other, they could be considered as indifferent.

In tabular form, the value of the net flows are shown in **Table C15**:

Table C15: Net flow values of case study example options

Alternative	Leaving (Φ^+)	Rank	Entering (Φ^-)	Rank	Net Flow (Φ)	Rank
In6S	0.3934	3	0.3352	4	0.0581	4
TBYF	0.5530	1	0.2302	1	0.3228	1
InSt	0.1616	6	0.5977	6	-0.4361	6
RSSC	0.4629	2	0.3332	3	0.1297	2
SeRe	0.3768	4	0.3139	2	0.0628	3
ROC	0.2986	5	0.4359	5	-0.1373	5

This illustrates that, given the performance matrix shown earlier and using PROMETHEE II, the 'Think Before You Flush' campaign is deemed to be the most feasible option in the sustainability of sanitary waste in the water/wastewater systems.

C4.3 Geometric Analysis for Interactive Aid (GAIA)

PROMETHEE can be extended graphically with the use of GAIA, which reduces the multi-dimensional criteria space to a two-dimensional criteria plane to allow direct visual presentation. It uses principal components analysis to project options described by their uni-criterion net-flows, criteria and the decision axis, defined by the vector of weights, onto a plane. From the GAIA plane, alternatives are represented by points, and criteria by axes. Criteria expressing similar preferences on the data are oriented in the same direction, whilst conflicting criteria are pointing in opposite directions. It is also possible to appreciate clearly the quality of the alternatives with respect to the different criteria. In addition to the representation of the alternatives and criteria, the projection of the weight vectors in the GAIA plane corresponds to another axis that shows the direction of the compromise resulting from the weights allocated to the criteria. When the weights are modified, the positions of the alternatives and the criteria remain the same, only the decision axis is changing. The movements of the decision stick corresponding to modifications of the weights are directly displayed in the 3D-view window – this illustrates the sensitivity analysis of the weights.

The GAIA plane, shown below for the case example, provides a comprehensive view of the decision problem:

- Criteria (or grouped categories of criteria) are represented by axes. By looking at the orientation of the axes, you can quickly identify the conflicts between the criteria.
- Actions are represented by their symbols. Their positions enable you to identify their strong and weak features.
- The Pi decision axis represents the weighing of the criteria. It shows you the type of compromise that corresponds to the current weights. You can modify the weights of the criteria and see how the Pi decision axis moves in the GAIA plane.

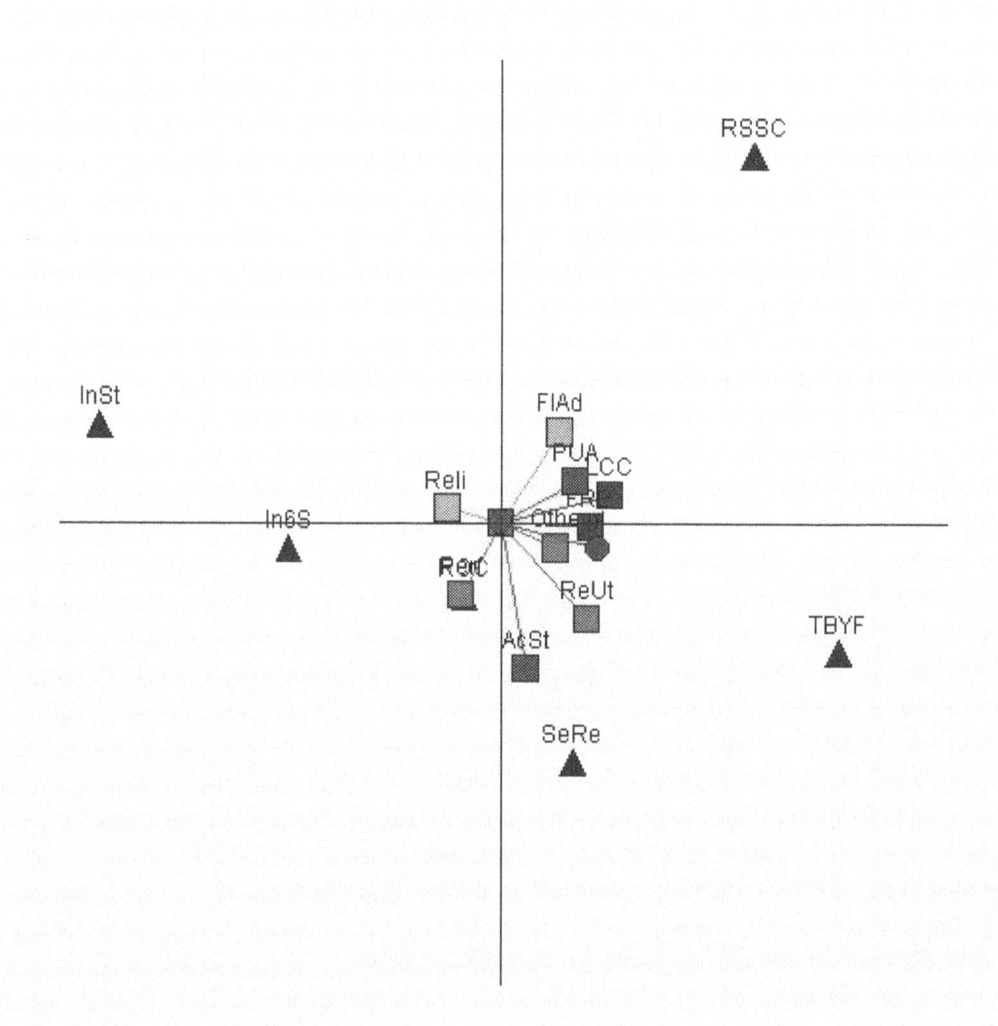

Figure C7: GAIA plane of case study example

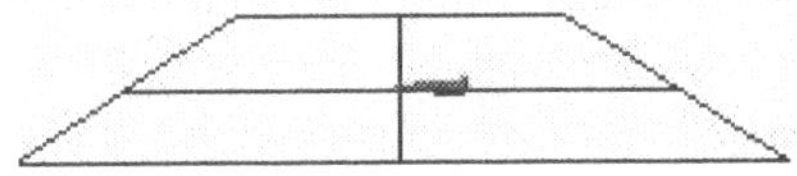

Figure C8: Decision stick of GAIA plane

C5 MULTI-ATTRIBUTE UTILITY THEORY (MAUT)

The MAUT method, developed by Keeney and Raifa (1976) in the mid 1970s, is based on the construction of a decision-maker's utility function in order to represent his/her preference structure, and provides assistance in choosing between a range of options.

The shape of the single attribute utility function curve is determined by the risk attitude of the decision-maker (see **Figure C9**). For a risk-averse decision-maker, the utility curve is curved and opening downward (concave). A risk-neutral decision-maker has a straight-line utility curve. And, for a risk-seeker, the utility curve is curved and opening upward.

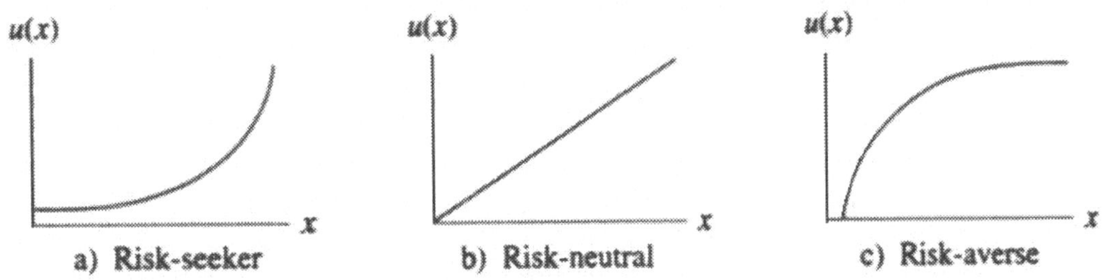

Figure C9: Shapes of utility curves

Some of the advantages of the MAUT technique are that it can explicitly distinguish the technical from the value components of a decision (i.e. the experts or the literature can provide technical information, whereas the decision-maker can assess the value judgements), it asks societal values which are closely connected with the problem, as well as it is very flexible in changing circumstances since it elicits broader range of values for each attribute. Also, MAUT presents, as an outcome, a range of feasible alternatives rather than one best solution – an objective preferred by some decision-makers. MAUT does also have the advantage that it has the ability to explicitly take into account the uncertainty in the decision outcomes (i.e. in many decision situations, it is not possible to determine which course of action would lead to the best expected outcome).

Also, MAUT does not need holistic judgements, and thus can accommodate the multi-dimensionality of value and allow the decision-maker to evaluate several alternatives with different environmental impacts in terms of minimisation of adverse environmental impacts.

MAUT has been criticised by many on the grounds of being too complex and requiring judgements that are difficult to elicit from decision makers, being based on unrealistic assumptions or simply being an unrealistic concept. On the back of this criticism, Edwards and Newman developed the simpler SMART method in the early 1980s.

C6 SIMPLE MULTIPLE ATTRIBUTE RATING TECHNIQUE (SMART)

SMART is a simplified version of the MAUT method that was specifically designed for application to social decision making. Ratings of alternatives are assigned directly in the natural scales of the criteria. In order to keep the weighting of criteria and rating of alternatives as separate as possible, the different scales of criteria need to be converted to a common internal scale. This is done mathematically by the decision-maker by means of a

"Value Function", in which one of the simplest choices, and in most cases is sufficient, is a linear function.

One of the advantages of using the SMART method is that the decision model is independent of the alternatives. While the introduction of value functions complexifies the decision-making process, the advantage is that the ratings of alternatives are not relative, so that in changing the number of alternatives considered will not in itself change the decision scores of the original alternatives.

The SMART procedure comprises three stages:

1. Prioritisation - as to which from a range of options for sustainable development
2. Evaluation of the options – namely sensitivity analysis
3. Monitoring and feedback to review each decision application to stage 1 and 2

C6.1 SMART case example

The basic procedure for the SMART method is now described. SMART is started by constructing a value tree, by addressing the attributes that are of general concern. These are broken down into attributes associated with these general attributes, and continued until all attributes have been recognised. The value tree for the current case study can be seen in **Figure C10.**

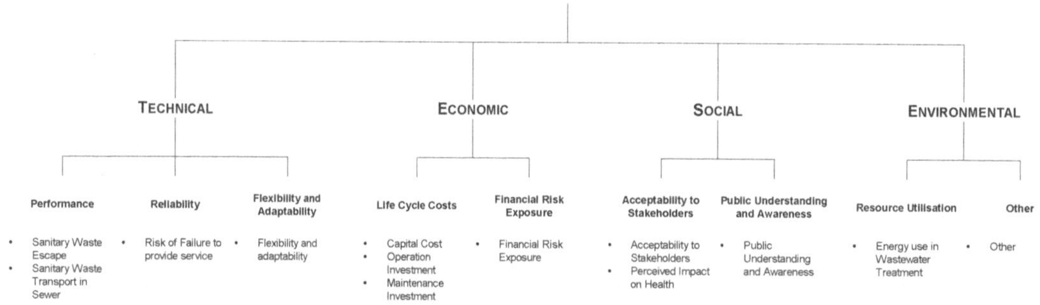

Figure C10: Value Tree of Criteria

Having constructed a value tree, it is necessary to judge whether it is an accurate and useful representation of the decision-maker's concerns. Keeney and Raiffa (1976) have suggested five criteria that can be used to judge the tree:

1. Completeness: All attributes which are of concern to the decision maker have been included;
2. Operationality: All lowest level attributes are specific enough for the decision maker to evaluate and compare for the different options;
3. Decomposability: The performance of an option on one attribute can be judged independently of its performance on other attributes;
4. Absence of redundancy;
5. Minimum size.

Firstly, the ranking and weighting of the criteria was created through decision-maker participation and discussion.

The procedure to rank the criteria weight within four categories was consolidated into group of criteria between inter-group and intra-group weightings. In order to do the analysis, the average weights value of criteria is formulated: The generic group weighting was multiplied by the individual criteria weight, normalised (by dividing each weight by the sum of the overall weight), and then multiplied by 100. This gave each criteria a weighting factor relative to its important within the context of the criterion as a whole, and is illustrated in the following table:

Table C16: Weighting factors of case study example criteria

	Weight
Perf	8
Reli	9
FlAd	7
LCC	10.5
FRE	3.5
AcSt	4.5
PUA	3
ReUt	3.5
Othe	3.5

(Note: that the sum of the weighting factors do NOT add to 100 as these weights were extracted from the case study)

Next, the value functions were developed for each criterion. A performance score of 100 was allocated to the criteria operating at maximum capacity, and a score of 0 to the minimum. The use of 0 and 100 as the least and most preferred values made the judgements easier and arithmetic simpler. The performances of each of the criteria, with respect to each of the options (Table C18), are converted to this direct rating scaling. The direct rating scale for each of the criteria is illustrated in the following table:

Table C17: Direct rating scales of case study example criteria

	In6S	TBYF	InSt	RSSC	SeRe	ROC
Perf	95	75	60	40	50	90
Reli	100	25	85	60	85	0
FlAd	60	60	0	100	0	0
LCC	45	80	40	95	65	70
FRE	70	100	0	100	100	0
AcSt	69	86	63	52	89	66
PUA	0	100	0	100	0	100
ReUt	25	95	0	30	90	50
Othe	80	95	70	85	90	95

Next, the weighted sum method is implemented on each of the criteria. The weighted sum method is a straightforward score analysis method, where the average weightings are multiplied by the performance score. Then all the criteria for each option is summed and divided by 100 leading to the overall performance of each option. This is illustrated in the following table:

The results establish that the RSSC option is the top ranked, but only marginally better than the TBYF option.

Table C18: Performance scored of case study example options

	In6S	TBYF	InSt	RSSC	SeRe	ROC
Perf	760	600	480	320	400	720
Reli	900	225	765	540	765	0
FlAd	420	420	0	700	0	0
LCC	472.5	840	420	997.5	682.5	735
FRE	245	350	0	350	350	0
AcSt	310.5	387	283.5	234	400.5	297
PUA	0	300	0	300	0	300
ReUt	87.5	332.5	0	105	315	175
Othe	280	332.5	245	297.5	315	332.5
Score	34.755	37.87	21.935	38.44	32.28	25.595

C7 RISK MANAGEMENT IN THE DECISION MAKING PROCESS

Risk can be defined as a combination of the probability, or frequency, of occurrence of a defined hazard and the magnitude of the consequences of the occurrence. A hazard is commonly defined as a property or situation that, in particular circumstances, could lead to harm. Generic risk assessment processes involve risk estimation and risk evaluation, where risk estimation is concerned with establishing the probability of occurrence of a risk arising from outcomes of a decision or action and risk evaluation is concerned with evaluating the impact of the risk on those affected.

Risk analysis methodologies adopt qualitative and quantitative approaches and the latter require the application of complex, data intensive modelling approaches. Such approaches are not considered in the SWARD processes because they are not designed to facilitate the direct impact of individual risks arising from the options under consideration. Risk assessment in the context of the SWARD processes is focussed on the selection and application of methods that can be used with the analytical tools to guide the decision-maker on the likelihood of making a wrong decision i.e. the risk of not identifying the more sustainable option.

This risk arises due to variability and uncertainty within the data supporting the analysis. Some data used in the decision analysis will have been based on estimates developed using human judgement or from modelling approaches that will have a margin of error. This uncertainty can undermine the reliability of the decision analysis. Two appropriate risk analysis methods that can be used with the analytical tools to guide the decision-maker on the risk of reaching a wrong decision are outlined here. These are sensitivity analysis and Monte-Carlo simulation.

Sensitivity analysis involves making a change to the value of one variable in the analytical model, whilst retaining the original value of the other variables. In this way, the influence of that variable on the decision can be evaluated. The robustness of the decision can be tested under different scenarios by, for example, considering best and worst-case assumptions for key variables. If the ranking changes little (insensitive to changes) then the risk of reaching a wrong decision is low. If the ranking changes significantly (sensitive to changes) then the risk of reaching, a wrong decision is a high. This process must be repeated for each of the key variables.

The overall results of the sensitivity analysis can be presented in tabular form or as a spider diagram. A drawback of sensitivity analysis is that it can only examine the influence of changing a single variable at a time.

Monte-Carlo simulation allows the effects of simultaneous changes in all the variables in the analysis to be observed. This is achieved by representing the parameters in the analysis by appropriate probability distributions, and then sampling and combining values of these parameter in an iterative process to produce a probability distribution of the results of the analysis (e.g. range of possible SMART scores for each option).

C8 VALUE MANAGEMENT (VM)

VM has already been described (in Section 2.5 of the guidebook) as:
"A systematic, multi-disciplinary effort directed toward analysing the functions of projects for the purpose of achieving the best value at the lowest overall life cycle cost." (Norton & McElligott, 1995)

C8.1 VM Evaluation Methods

As described in Section 2.5, the more common "evaluation techniques" fall under three categories:

1. Selection by Value Management Team Co-ordinator (VMTC)
2. Simple democratic
3. Complex democratic

It should be pointed out that VMTC is simply another term for Value Specialist or VM Facilitator.

1. Selection by VMTC

The following method may be described as the 'benevolent autocratic' approach.

In this case the VMTC selects the ideas for development but asks the others if they agree with the selections. Team members thus have a limited input in their editing role. This technique may be fast but, in restricting discussion, the identification of all relevant criteria may not occur. Accordingly, a poorer quality of selected ideas may result. This technique should therefore only be used when there are strict time constraints and/or the team members are insufficiently experienced.

2. Simple Democratic

A simple democratic selection process may be achieved by use of either consensus rating or voting techniques.

The consensus rating technique requires team members to discuss the ideas and agree on an overall rating for each idea. Several rating schemes can be applied, one example being the numerical 1-4 scheme. In this scheme the numbers represent the way the idea should be treated (see below):

1. Idea is to be developed
2. Idea is to be developed if there is sufficient time
3. Idea has some merit but is not to be developed
4. Idea has little merit and should not be considered.

This single rating scheme is said to be very effective. Dual rating is a variation of this scheme where an initial technical/functional rating is combined with a cost impact rating. However, dual rating may be more time consuming than single rating. Furthermore, it can eliminate good technical ideas on cost grounds even though they represent good value.

Ideas may also be rated by simply accepting or rejecting them. This scheme, in lacking compromise positions, means that arguments for acceptance and rejection are more fervent. Thus the scheme may be time consuming, as well as failing to provide relative scoring.

The final 'simple democratic' technique is voting. Team members vote for ideas that they believe have merit. Each member is awarded a limited number of votes – normally half or less of the number of available ideas. Members are allowed to cast only one vote for any individual idea and must vote independently. The numbers of votes for each idea are totalled and the team (or VMTC) decides how many votes are needed for an idea to be developed. It is recommended that the voting results be discussed in order to highlight ideas that have been voted as favourable but, in reality, are impractical for reasons not initially understood by those voting for them.

3. Complex Democratic

The term "complex democratic" is used by Norton & McElligot (1995) to describe techniques that "first weigh evaluation criteria according to their relative importance and then score alternatives on the basis of the weighted criteria to determine those that are optimal". These techniques have also been described as "weighted preference methods" by Green (1992).

Norton & McElligott recommend a specific technique, which is a combination of the "Criteria Scoring Matrix" and "Alternative Analysis Matrix".

The criteria scoring matrix is used to establish relative weights for each criterion. This is achieved by comparing the importance of each criterion with that of every other criterion. The more important criterion is scored as follows:

4 Major preference over other criterion
3 Medium preference over other criterion
2 Minor preference over other criterion
1 No preference over other criterion

The scores for each criterion are totalled to provide a relative weighting. For convenience, the raw weighting scores are converted into a range of 1-10. Criteria with a raw score of zero are assigned weightings of 1 so they are not eliminated totally from consideration during the comparison of alternatives.

The alternative analysis matrix is then used to score each design alternative against each criterion as follows:

5 Excellent
4 Very good
3 Good
2 Fair
1 Poor

The scores of alternatives against each criterion are multiplied by the weighting for that criterion. Finally, the weighted scores for each alternative are totalled and the alternative with the highest score selected.

Norton & McElligott admit that "the major benefit of the technique is not from quantitatively calculating the 'best' solution, but is derived from being forced to review all the pertinent criteria in an objective manner".

C8.2 Criticisms of VM

Some VM practitioners have claimed that VM is unique because "it involves analysing the functions of a project" (Norton & McElligott, 1995). Others claim that "there is no alternative management tool or technique available that can be used for the purposes of VM, or to get the benefits gained from a rigorous VM process" (Best & De Valence, 1999).

However, Green (1992) has argued that VM does not appear to possess any unique characteristics. The VM 'job plan', for example, can be summarised as:

- analyse nature of problem;
- generate alternative solutions;
- select best option.

Green notes that this staged approach is not unique to VM and that several methods of creative problem solving exist which are based on similar methodologies.

He also likens a VM study to a decision conference, an approach already well described in decision theory literature. The characterisation of a decision conference as quoted by Green (1992) reveals the similarities.
"*...the use of a structured process for dialogue and debate amongst the decision maker and his experts during an intense short-term conference...The objective of a decision conference is to focus attention on the problem in hand, to identify and evaluate a set of meaningful and practical options, and to synthesise a group consensus in favour of a single option*"

Furthermore, he compares the 'how-why' logic of FAST diagrams (a VM functional analysis technique) with the 'means-ends' analysis of objectives (not VM specific) and finds that the approaches are essentially the same.

Green concludes that "in many respects, the term 'value management' is merely a convenient label for a multi-disciplined decision conference which follows a structured format".

While there may not be anything intrinsically 'wrong' with VM, it appears not to be as distinctive as some practitioners would claim.

It was suggested earlier that VM may improve the inclusion of sustainability in decision making by:

- promoting stakeholder inclusion;
- concentrating on the achievement of best value;
- allowing for the consideration of social and environmental (as well as technical and economic) functions;
- emphasising life-cycle costs (as opposed to only capital costs).

If the decision process does actually conform to the above ideal, then the outcome will almost certainly be more sustainable than that from a process which does not. In other words, VM can improve the inclusion of sustainability in decision making – if the claims made for it can by realised in practice.

In the real world of decision making, there are numerous factors that could reduce the effectiveness of the VM process.

One such factor is the size of the workshop team. Norton & McElligott recommend six to twelve participants, as larger teams can have the following negative consequences:

- team can become relatively difficult to control;
- large teams tend to inhibit some members, hindering full team participation;
- establishing a good rapport and team spirit is more difficult, thus making it harder to maintain enthusiasm and productivity throughout the workshop.

However, workshop teams of 20 or more are not unheard of, presumably at the client's request.

Another factor is the composition of the workshop team. Although key stakeholder groups should be identified and included, it is by no means certain that this will occur.

The duration of the workshop may also have an influence on it effectiveness. If a client agrees a study duration with a VM specialist, the specialist may feel obliged to 'get some decisions' within this timeframe, even if these are not the 'best value' decisions.

REFERENCES

Best, R. and De Valence, G. (1999), Building in value, Arnold, London, ISBN 0 340 74160 0.

Brans, J.P. and Mareschal, B. (1990), The PROMETHEE method for MCDM; The PROMCALC, GAIA and BANKADVISER software. In: Bana e Costa C. (Ed.) Readings in Multiple Criteria Decision Aid. Springer-Verlag, pp. 216-252.

Brans, J.P. and Mareschal, B. (1994), How to decide with PROMETHEE. [online]. Available as a white paper from: http://www.visualdecision.com/download.htm [Accessed 18/07/00].

Brans, J.P. and Vincke, P. (1985), A preference ranking organisation method (the PROMETHEE method for multiple criteria decision-making). Management Sci. 31(6), pp. 647-656.

Brans, J.P., Mareschal, B., and Vincke, P. (1984), PROMETHEE: A new family of outranking methods in multicriteria analysis. In: Brans JP. (Ed.). Operational Research'84. Elsevier, North Holland, Amsterdam, pp. 408-421.

Edwards, W. and Newman, J.R. (1982), Multiattribute Evaluation. Sage, London, ISBN 0 8039 0095 3.

Green, S.D. (1992), A SMART Methodology for value management, Englemere: The Chartered Institute of Building, ISBN 1 85380 055 4, ISSN 0306 6878.

Keeney, R.L., Raiffa, H. and Meyer, R. (1993), Decisions with Multiple Objectives, Preferences and Value Tradeoffs. Cambridge University Press, Cambridge, ISBN 0 5214 3883 7.

Norton, B.R. and McElligot, W.C. (1995), Value management – a practical guide, MacMillan, London, ISBN 0 333 60626 4.

Roy, B. (1991), The outranking approach and the foundations of ELECTRE methods. Theory and Decision 31(1), pp. 49-73.

Saaty, T.L. (1980), The Analytic Hierarchy Process. McGraw-Hill, New York, ISBN 0 0705 4371 2.

Appendix D1: Water Supply, Treatment and Distribution Strategy

Contents

D1.1. INTRODUCTION

This Case Study concerns the efforts of one of the Scottish Water Authorities (now Scottish Water) to develop an "integrated strategy for water supply, treatment and distribution" to enable it to "meet current and future demands up to 2020" and "respond effectively to competition".

Although this Water Supply, Treatment and Distribution Strategy was being developed for a specific geographical area, it was intended to serve as a model for strategies in other areas. The Authority's approach to the development of the Strategy involved the use of Value Management, using an external facilitator, and involving a variety of internal/external stakeholders including a number of external consultants. A comparison of this approach with that developed within the SWARD project was therefore of great interest.

The main aims of the Case Study were to:

- describe the processes and tools used in the development of the Strategy;
- show how sustainability has been taken to account in the Strategy;
- compare the Strategy processes and tools with those of the SWARD guidebook.

Research was performed in close collaboration with this particular Authority on a series of studies throughout the SWARD project. This particular Case Study was the third Value Management exercise we have observed within the Authority and by far the most high-level. The first concerned the determination of the Authority's sludge management strategy. The second concerned water distribution/network decisions relating to security of supply/levels of service within part of a major city and where environmental and local stakeholder concerns were a major factor.

A key outcome of these observations was the development of the Decision-Mapping (Bouchart et al., 1999 and 2002) approach that provided the SWARD research team with an invaluable insight as to how real decisions were made within WSPs. It enabled us to see exactly how the decision process was structured, how the appropriate information was assembled and used, how decision priorities were determined and where gaps in the process/data might occur. More specifically, we were able to pinpoint the extent to which sustainability considerations were considered. The results of the earlier decision-mapping exercises were fed back to the WSP's and helped in part to frame the prioritisation approach taken in this Case Study. Equally, the decision-mapping research contributed significantly to the way the SWARD project evolved and in particular to the general structure and tone of the SWARD process as described in the Guidebook.

As will be seen later, the Strategy described in this Case Study differs from the subjects of other SWARD case studies in terms of its scale, complication and complexity. The Strategy has to deal with many water supply assets in a relatively large geographical area, including lochs, reservoirs, boreholes, treatment works, trunk mains and distribution networks. Trying to determine the best strategy for the operation, upgrading, or replacement of these assets is a problem that is both complicated ("rich in detail") but also complex ("rich in structure"). Besides 'hard' analytical processes, there is therefore a need for 'soft' processes for effective communication, negotiation, discussion and decision making.

For reasons of commercial confidentiality it is not possible to disclose the identity of the Authority, the names of those involved in the Strategy, or the location of the Strategy area. Some technical aspects have likewise been modified or described in very general terms. However, as the case study is primarily concerned with understanding the Authority's decision-making processes, a lack of technical detail should be of little consequence.

Although the SWARD guidebook is not actually applied in this Study, it was felt that an account of a WSP's real experience of using a 'guidebook-type' process would be invaluable in helping to benchmark the SWARD approach.

D1.2. CASE STUDY BACKGROUND

D1.2.1. Strategy area and history

The Strategy covers an area of over 1,500 sq. km with a wide topographical variation. Water is primarily supplied from upland surface water sources with groundwater sources providing support during the summer months. While the majority of water resources are located in the west, most demand is in the centre and east. This has led to the development of long supply lines operating at high pressures and a distribution system characterised by a large number of smaller service reservoirs, break pressure tanks and pressure control zones.

The area has about 350,000 customers and is supplied with around 160 Ml of water per day. In average years, the area enjoys a significant water supply surplus but in drought years may be subject to deficit. It has been estimated that, in 2020, a 1 in 50 year yield would result in a supply/demand deficit of 77 Ml/d. This deficit increases to 88 Ml/d for a 1 in 100 year yield.

In 1995 a request was made at short notice for very large amounts of water to service a new industrial development. This, together with the possible water deficit in drought years, resulted in the 1996 Augmentation Study. It concluded that a phased delivery of additional water resources from an adjacent area was the only solution. The cost of Phase 1 of this project was estimated at nearly £20 million.

The industrial development has not proceeded as expected and, even now, uncertainty exists as to when this demand will occur. This uncertainty, coupled with the high cost of the augmentation project, has highlighted the possibility of deferring this major investment while progressing a range of less expensive measures under an area water strategy.

D1.2.2. Objectives and key issues

The objectives of the Strategy were agreed in January 2001 and were:

- to optimise operational practices across water supply, treatment and distribution to achieve the best value between whole life cost, operational risk, income opportunity and sustainability.
- to optimise capital investment requirements in terms of whole life cost analysis and to provide an integrated, auditable approach to decision making.

The key issues of the Strategy centred on understanding:

- source yields and potential deficits;
- actual and economic leakage levels;
- the proposed 20 year demand forecast;
- major risks and potential mitigation;
- the network investment backlog;
- capital investment needs, priorities and consequences of non investment.

D1.2.3. Outputs

The final outputs of the Strategy will be a Strategy Document (for this particular area) and a Strategy Document Model (for use in future strategies).

The Strategy Document is required to provide a "fully referenced and auditable set of data models, which are also to be delivered in software form, capable of being revised by new input data, to allow the Document to be truly dynamic."

D1.3. DEVELOPMENT OF THE STRATEGY

D1.3.1. Organisation

The structure of the strategy project team, and some of its responsibilities, is shown below.

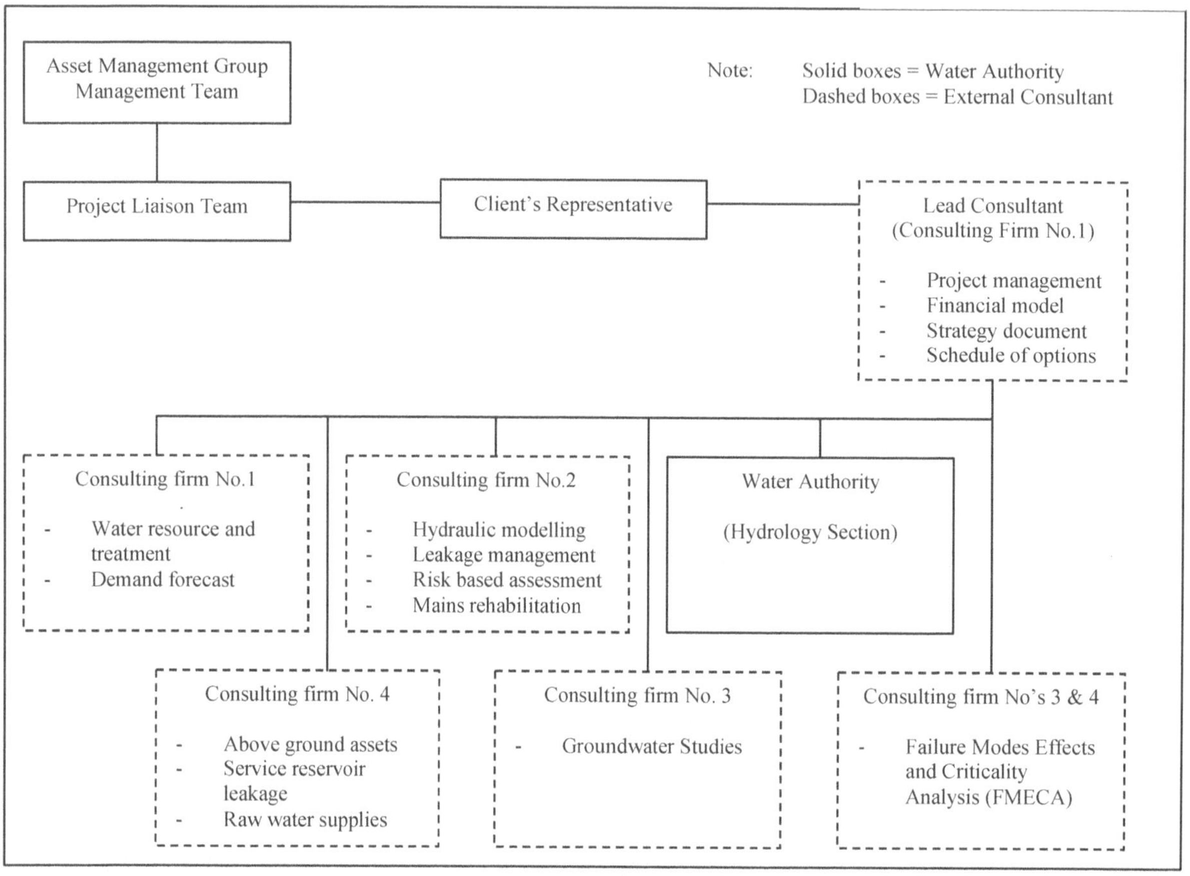

Figure D1.1: Project team structure

It can be seen from **Figure D1.1** that the lead consultant provides a single point of contact for the Water Authority (the client). The lead consultant is responsible for the management and co-ordination of the various supporting activities being carried out by the Authority and its external consultants.

In the production a single Strategy document, the lead consultant has to bring together the outputs of some 15 related projects/studies involving a total of 33 people. Of these 'stakeholders' 18 are Authority employees; 14 are outside consultants and 1 is from a neighbouring water authority. Each stakeholder has a defined role and responsibility. Typical roles include "project sponsor", "project team member" and "consultant", with corresponding responsibilities of "delivery of study", "operational input" and "delivery of report on works".

The documentation relating to the Strategy suggests that a formal 'project management' approach has been used, in line with accepted best practice. Evidence to support this includes the early agreement of Strategy purpose and objectives; the production of a project plan and risk register; and the use of regular progress meetings.

D1.3.2. Timescale (from June 02, May 02)

Figure D1.2 illustrates the phases of the strategy and some of the kev events.

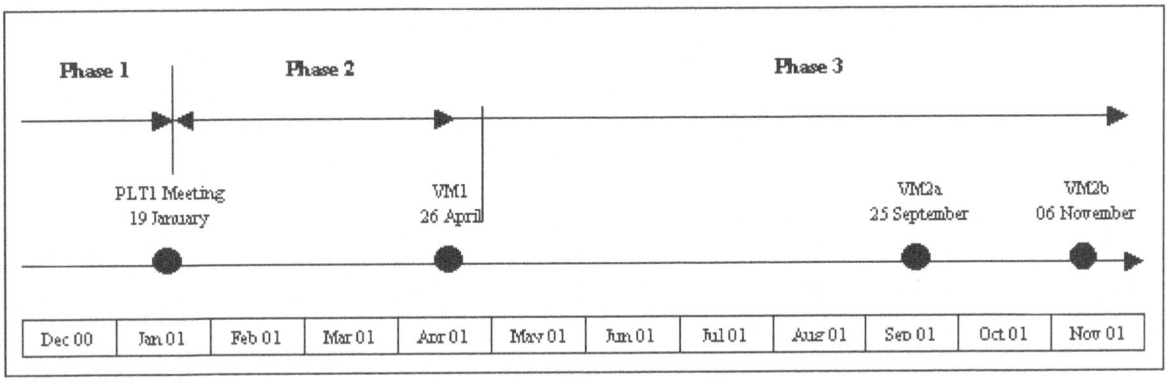

Figure D1.2: Strategy timescale

As can be seen, the Strategy has developed over a period of 12 months and can be divided into three distinct phases. Each of these phases is related to specific events, in particular the first Project Liaison Team meeting (PLT1) and the three Value Management Workshops (VM1, VM2a & VM2b). Note that VM2b is simply a continuation of VM2a.

Due to time constraints, the cut-off point for this Case Study is the VM2b workshop, held on 6 November 2001. The Strategy Document and Strategy Document Model should be handed over by mid November 2001. This marks the end of 'strategy production'. The next step is to actually implement the strategy and monitor its results over a 20-year period.

D1.3.3. Strategy phases

Phase 1 appears to have been used to prepare the way for the rest of the Strategy. It concluded with the first PLT meeting, the purpose of which was to:

- reaffirm the project team structure, Strategy purpose and objectives;
- review the Strategy's current state of development;
- agree a Strategy plan;
- review the Strategy programme in detail.

The main purpose of **Phase 2** was to generate information for use in VM1. This included:

- determining water source yields;
- agreeing levels of service;
- determining the economic level of leakage;
- estimating demand;
- summarising supply/demand balance for different scenarios;
- collating asset condition and performance information;
- collating water quality and customer information;
- production of a first-pass range of short and long-term options;
- determining Opex and Capex figures for option appraisal

VM1, which will be described in more detail later, highlighted a number of points such as:

- the need for a review of the demand forecast;
- short term operational concerns (e.g.localised areas of network with frequent bursts);
- options to be further developed, prior to the next VM workshop.

Phase 3 is the last phase of the strategy development process. This phase consists of:

- preparation for VM2a and VM2b;

- the workshops themselves;
- the production of the Strategy document and model.

The Authority had originally planned to hold only one VM2 workshop. In fact, it was necessary to hold a 'main' workshop (VM2a) and a 'follow-up' workshop (VM2b).

By mid November 2001 - and the production of the Strategy document and model - Phase 3 will be virtually complete. However, a number of actions arose from VM2b that will have an impact on the Strategy. In effect, these actions will extend phase 3 into early 2002.

D1.3.4. Strategy options

The asset management options under consideration can be classified as either **capacity options** or **security options**.

Capacity options are those that would reduce or eliminate the risk of a water supply deficit during times of low yield.

A wide variety of capacity options were suggested, including:

- leakage reduction;
- upgrading of water treatment works;
- import of water from other areas;
- borehole refurbishment;
- development of new boreholes;
- conjunctive use of sources;
- river augmentation.

Security options are those that would reduce the risk of interruptions to customers' water supply. Such interruptions should not last more than 12 hours, according to agreed level of service criteria. Following a structured risk assessment of the water supply network (trunk mains and local distribution) and its storage (service reservoirs), a number of improvement schemes were proposed. These included:

- mains duplication;
- installation of cross connections;
- the provision of additional storage.

From the available documentation, it is evident that the Strategy has concentrated mainly on capacity options, with security options receiving less attention.

VM2a produced a preferred package of capacity options. From this, the preferred package of security options, discussed in VM2b, followed almost automatically.

D1.3.5. Strategy Processes/Tools

The precise meaning of terms such as process, procedure, support system, framework, approach, tool and methodology is often a source of confusion.

For this case study we are particularly interested in the VM decision-making process. Exactly how do VM participants, confronted with a range of options and an abundance of information, arrive at a decision?

In this context VM can be thought of as a procedural tool as opposed to an analytical tool. A procedural tool has been defined elsewhere in the guidebook as one which "focuses on procedures to guide the way to reach a decision" whereas an analytical tool provides "information as to the consequences of a choice". This is illustrated in **Figure D1.3.**

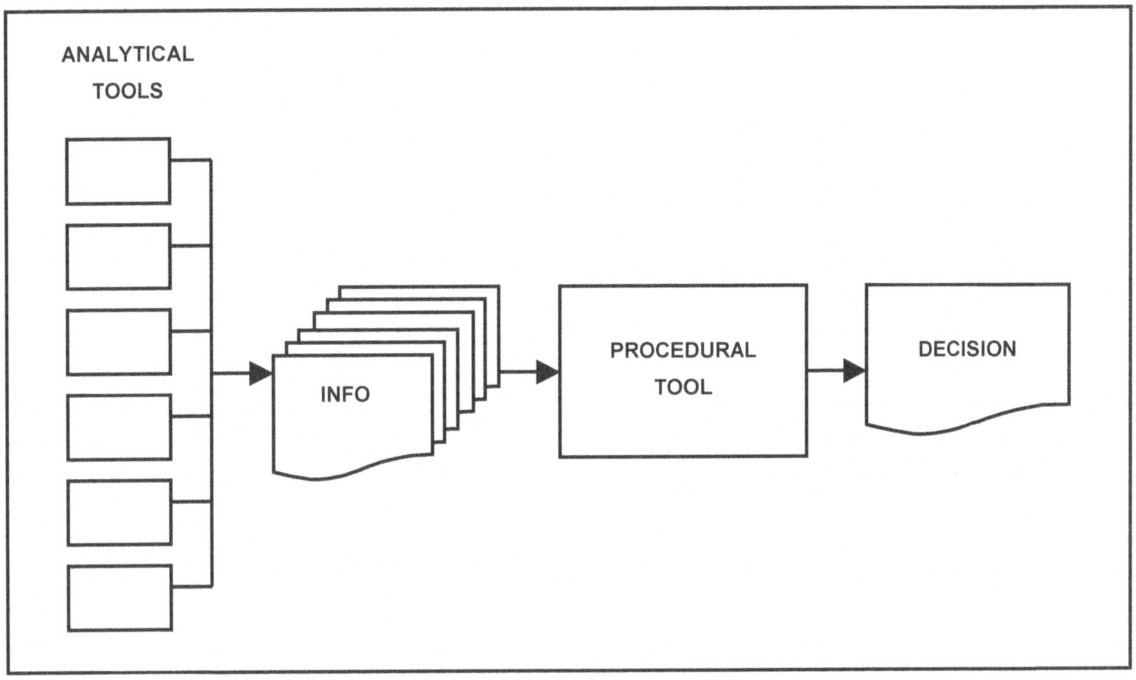

Figure D1.3: Analytical and Procedural Tools

Analytical Tools
In providing information as to the consequences of a decision, analytical tools must provide information about the current situation as well as information about possible future situations.

Numerous analytical tools have been used to help generate Strategy information. Most of this has been formalised into documents that comply with the Authority's standard format. These documents are available to Authority employees and are controlled by formal change procedures. For this study, we were fortunate to have been given the majority of the Strategy-related documents, dated from March to May 2001. These 17 documents contain a total of over 400 pages of information.

It would be impractical to attempt to identify every 'tool' used to provide the above information. Instead, some of the main analytical tools and their uses are given below:

HYSIM

- Specifically adapted for Scottish catchment hydrology.
- Used to determine accurate hydrological data for analysis of long term reliable yields under a variety of drought events.

AQUATOR

- Software suite used for conjunctive use modelling of the major surface water sources.
- Makes use of HYSIM yields

Economic Level of Leakage (ELL) Model

- Developed to assess the short run economic level of leakage for the Strategy area.
- Various data sources used for the model:
- Infrastructure data (main lengths, property counts, etc);
- Burst data (reported mains and communication pipe bursts);
- Marginal cost of water (derived from supply and treatment costs).

Demand Forecasting Model

- Developed to assess the impact that future changes in water usage will have on water demand in the Strategy area (over next 20 years).
- Follows, as far as is practicable, the UKWIR demand forecasting methodology.
- Various data sources used for the model, including:
- 1999-2000 Water treatment works output data;
- Predictive data from the Scottish Executive;
- Analysis of Water Authority databases.

Cluster Analysis (operational customer contacts and bursts)

- Used to locate mains that feed properties with a history of operational customer contacts.
- Operational customer contact cluster defined as: "Fifteen or more operational customer contacts per 50m of main".
- Operational contacts caused by discoloured water, no water, taste/odour problems, leakage, low pressure, etc.
- Also used to locate sections of mains prone to bursting.
- Burst cluster defined as: "Average of five or more mains bursts for 50m individual section of main".
- Cluster analysis uses data from the Authority's customer contact and works management databases.

Network Model

- Constructed to represent the strategic elements of the water supply and distribution system.
- Almost 500km of mains pipework included in model.
- Single model (created in INFOWORKS) derived from 8 hydraulically separate models (created in WESNET).
- Used to show current and future network performance.
- Identifies pipelines already operating at or near maximum capacity (headlosses greater than 5m/km).

Water Infrastructure Assets Risk Analysis

- Analysis identifies water mains that are in poor condition and also serve sensitive or large numbers of customers. These mains require urgent detailed investigation and potential replacement within two years.
- Analysis developed in accordance with the Water Industry Commissioner for Scotland's econometric model.
- Risk analysis comprises of two main assessments:
  - operational risk analysis;
  - financial risk analysis.
- The final risk classification is the worst case from either analysis.
- Both analysis types are based on a combination of Likelihood of Occurrence and Impact of Occurrence.
- Analysis uses corporate data and other predictive and analytical models.

Operational Performance Costing (OPC) Analysis

- Used to calculate current inferred marginal operational costs that are directly linked to particular groups of assets.
- Also used to model the impact of the following leakage management options:
- Active Leakage Control (ALC)
- ALC and Pressure Management (PM)
- Rehabilitation, ALC and PM.

- OPC analysis therefore used to predict the potential level of leakage saving if certain techniques are used.
- Leakage calculations based upon burst and background losses and frequency relationships, all of which are pressure related.

Environmental Assessment

Some options have been subject to an initial environmental scoping exercise, carried out in consultation with interested parties.

- Scoping exercise not intended to be an exhaustive study of the potential environmental impacts, but instead identify potential key constraints.
- Constraints should be subject to more detailed studies (for options that are to be taken forward).

The scoping exercise considered the following potential effects:

- physical changes to protected sites and habitats of protected species;
- changes in water quality in reservoirs;
- changes in flow regimes;
- transfer of species and disease;
- transfer of pollution incidents.

The analytical tools described above have been used to reduce the gap between what is known and what needs to be known in order to make a decision. **Figure D1.6** shows the reduction of this information 'gap' as the strategy has progressed.

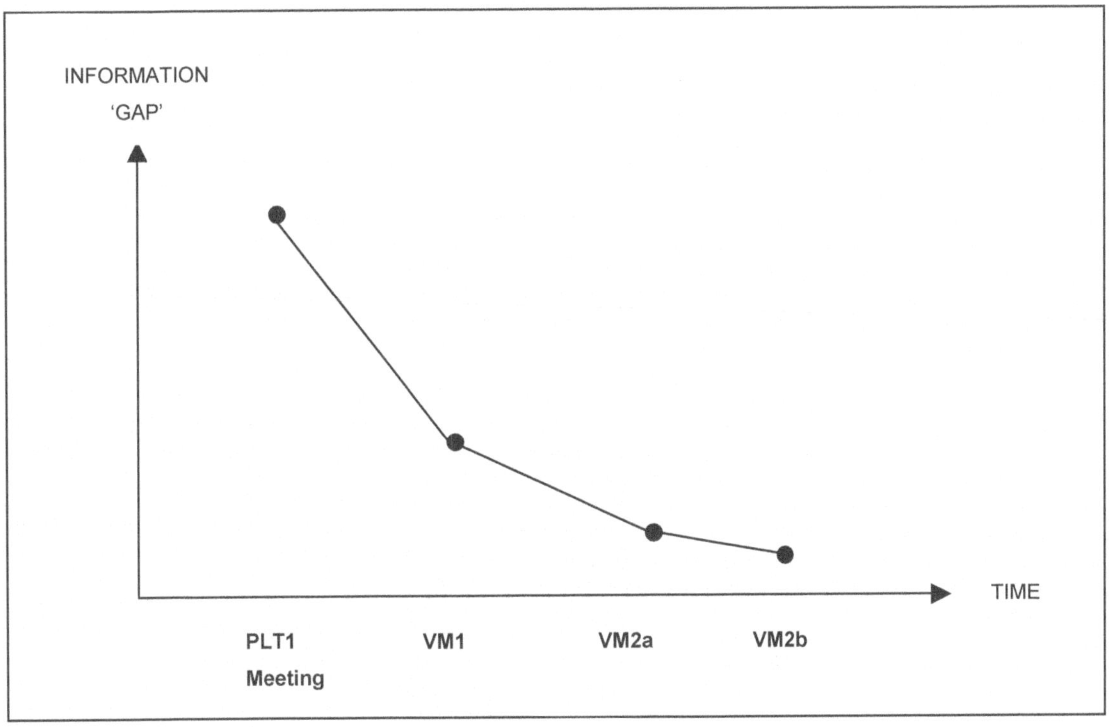

Figure D1.6: Reducing the information 'gap'

Following the first meeting of the project liaison team (PLT1), a large amount of information was generated for use in VM1. At this workshop some initial conclusions were reached regarding options for improved capacity and security. However, it was clear that these required further development prior to the next VM workshop. Furthermore, the accuracy of some VM1 information was questioned, the demand forecast being one example. Thus the information 'gap' had been reduced, but needed to be smaller still before any firm decisions could be taken.

By VM2a, the information was sufficient to allow a preferred package of capacity options to be decided. Even so, the actions from this workshop indicate a need for more information e.g.a reality check on the completeness of options and a review of detailed costings.

Between VM2a and VM2b, further analysis of capacity options resulted in a small, but worthwhile, improvement to the preferred package. VM2b itself was not really a decision-making workshop, as the security options discussed (and accepted) resulted from decisions taken at VM2a.

Value Management (VM): a Procedural Tool

Value Management has been briefly described in Section 2.5 and covered in more detail in Appendix C of the Guidebook. To recap, the guidebook definition of VM is:

"A systematic, multi-disciplinary effort directed toward analysing the functions of projects for the purpose of achieving the best value at the lowest overall life cycle cost."

A VM study normally has three main phases: pre-workshop, workshop and post-workshop.

1. Pre-workshop phase

This phase is concerned mainly with workshop organisation and provision of information to participants. The Client and Value Management Specialist will normally meet to establish study logistics, project issues and constraints. The VM Specialist will compile the relevant information (client brief, drawings, surveys, costs, etc) that is sent to participants prior to the workshop.

2. Workshop Phase

Workshops typically last for three to six days, depending on the project. Workshops follow a 'job-plan' consisting of the following phases:

Information phase:	Overview of VM process by Value Specialist. Presentations by design team. Functional analysis of project.
Creative phase:	Generate alternative ideas to achieve the same basic function. Rationale may be lower costs and/or improved functions. Brainstorming technique often used.
Evaluation phase:	Ideas from creative phase evaluated and best ideas selected. Economic and non-economic evaluation criteria used. Various evaluation methods may be employed.
Development phase:	Selected ideas developed into fully detailed proposals. Proposals include descriptions, pros/cons, initial costs, life-cycle costs, technical calculations, sketches, etc.
Presentation phase:	Proposal summaries presented to decision-makers. Draft copies of summaries handed over to decision-makers. Action plan written detailing follow-up work.

3. Post-workshop phase

A workshop report, containing a clear and comprehensive record of events and outcomes is prepared and circulated. Decision-makers are given an appropriate period to consider the recommendations of the VM team before an implementation meeting is held. The purpose of this meeting is to determine whether proposals should be accepted or rejected, and to establish subsequent actions.

Strategy VM Workshops
Of the three workshops held so far, only VM1 and VM2a can be considered 'decision-making workshops'. VM2b basically followed on from the decisions taken in VM2a. For this reason, and the fact that a workshop report is not yet available, VM2b will not included further in this section.

VM1 and VM2a can be described according to the following features:

Facilitator
The same VM facilitator, from an independent programme management company, led both workshops. The facilitator had been used on a number of previous occasions by the Authority and was probably known to most of the workshop participants.

Duration
Each workshop lasted for one day. In both cases a preliminary meeting was held about three weeks before the actual workshop.

Number of attendees
The numbers of workshop attendees are shown in **Table D1.1** below.

Table D1.1. VM Workshop attendees

	VM1	VM2a
Water Authority Employees	17	11
External Consultants	6	4
VM Facilitator	1	1
VM Facilitator's Assistant	1	1
Heriot-Watt University Observers	0	2
Total	25	19

Workshop Structure
The structure of each workshop was a variation on the generic VM 'job-plan' described earlier. This is shown in Table D1.2 below.

Table D1.2: VM Workshop job plans

Phase	Generic VM Job Plan	VM1 Job Plan	VM2a Job Plan
1	Information	Information	Information
2	Creative	Function	Function
3	Evaluation	Creative	Creative & Evaluation
4	Development	Evaluation	Development
5	Presentation	Development	Conclusions
6		Conclusions	Implementation
7		Implementation	

In both VM1 and VM2a the workshop information was presented by Authority employees and external consultants. Seven presentations were given in VM1 and four in VM2a.

A Value Tree was used during the function phase of VM1 to show the strategy objectives (see **Figure D1.4**).

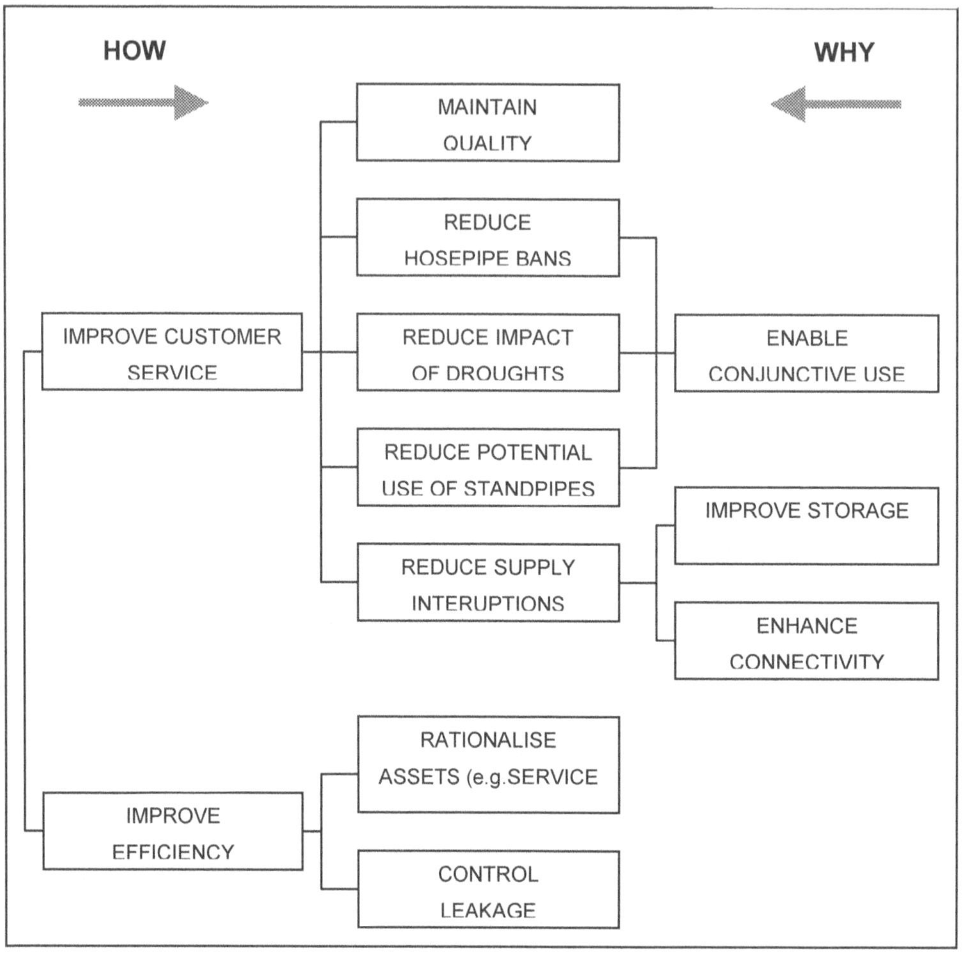

Figure D1.4: VM1 Value Tree

The technique used to produce the Value Tree is also known as FAST (Function Analysis Systems Technique). The resulting diagram represents project functions, their hierarchy and logical how-why relationships.

Unlike VM1, the function phase of VM2a was used to 'structure' the decision making process. The outcome of this phase was a decision-making process 'flowchart' (see **Figure D1.5**).

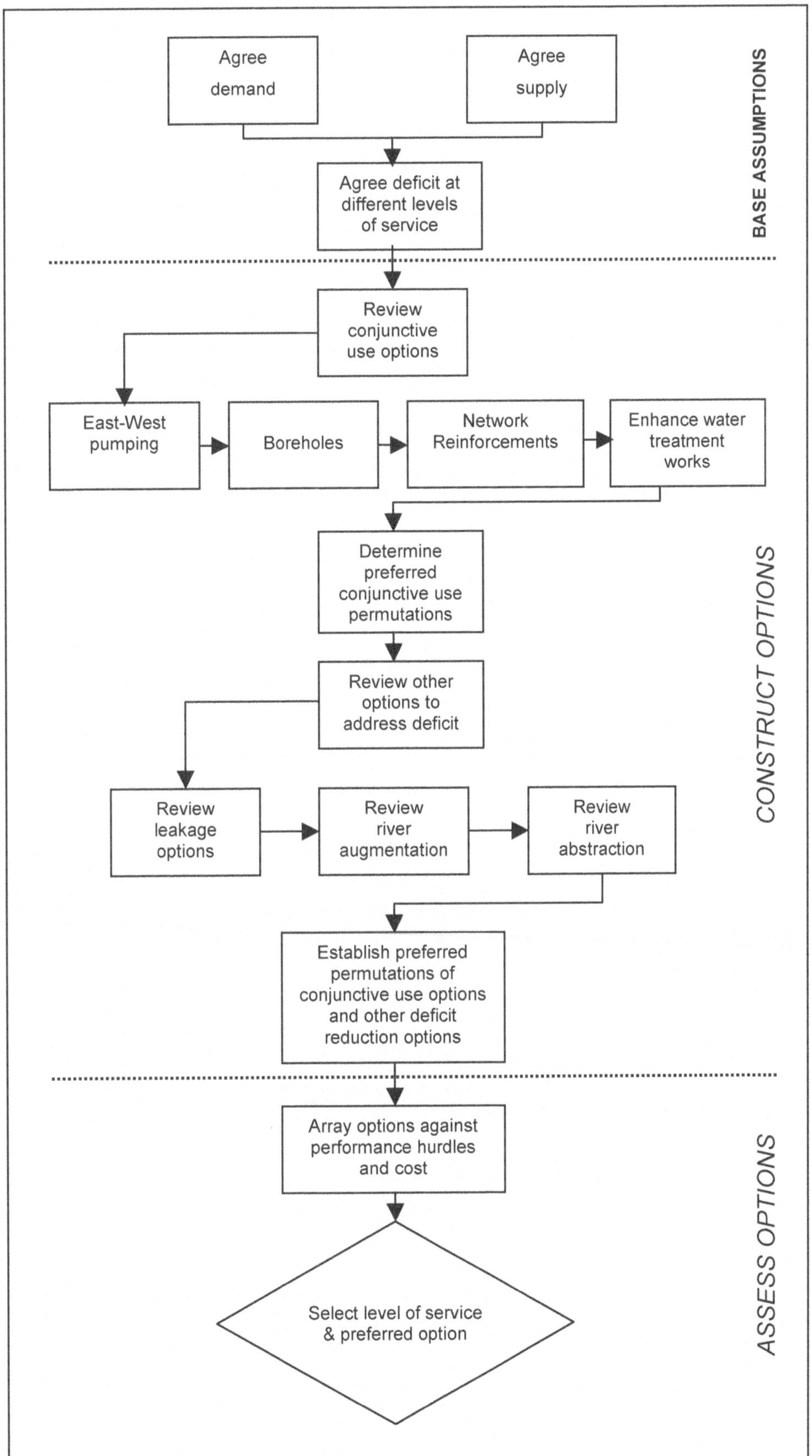

Figure D1.5: VM2a Decision Making Process

The creative and evaluation phases were separate in VM1 but combined in VM2a. Critical issues and potential options were listed in VM1 while only critical issues were listed in VM2a. In each workshop these items were prioritised for discussion by assigning a code of 1, 2 or 3, the meanings of which are given in **Table D1.3** below.

Table D1.3: Prioritisation codes

Code	VM1	VM2a
1	Discuss	Discuss/develop at workshop
2	Include in study	Action to be taken after workshop
3	Well understood, reject	Well understood, no further action required

The development phases of VM1 and VM2a were essentially the same. Critical issues and potential options were discussed individually and conclusions reached for each.

The conclusions phase of VM1 appears to have involved no more than the listing of workshop "key conclusions". This included operational concerns, options to be developed and further information required. The conclusions phase of VM2a differed in that it was actually used to evaluate options. Options were first assessed according to how well they met a variety of 'performance hurdles'. Options were then assessed according to their costs. From this process, a preferred option emerged.

For both workshops the implementation phase consisted of drafting an 'action plan'. This was a list of actions (tasks) to which responsible individuals were assigned together with estimated completion dates.

D1.3.6 Strategy Outputs

The Strategy has produced a package of preferred options that should ensure that, in the event of a 1 in 50 year yield, there is little or no supply/demand deficit. These options consist of:

- leakage reduction, achieved by active leakage control and pressure management;
- maximising use of existing water import facilities;
- upgrading of two water treatment works;
- refurbishment of existing boreholes to secure maximum output;
- develop new boreholes;
- reinforcement of key network mains;
- conjunctive use of water sources.

These options will be included in the Strategy Document for this area. A Strategy Document Model, for use in future strategies, is also being produced.

The Strategy has also highlighted a number of business policy issues that require consideration by senior management. These are principally:

- the levels of service to be adopted across the business;
- the principles of future demand planning;
- the adoption of a risk based approach to investment planning such as Failure Mode Effect and Criticality Analysis (FMECA).

Effectiveness of Strategy Processes/Tools

In meeting its objectives, the strategy development process *has* been effective. Decision-making was aided by the information generated by a comprehensive array of analytical tools. The key procedural tool used, Value Management, was also used to good effect.

Interestingly, much of the work that would normally be carried out during a VM workshop was actually done outside of the Strategy workshops. For example, during the creative phase it is normally the workshop team that generates alternative options for fulfilling the same basic function(s). However, this was not the case at the Strategy VM workshops. Instead, the alternative options were generated beforehand by another team. The Strategy VM workshop team therefore assumed a review function rather than a creative function. This can probably be attributed to the complication of the problem (over 190 combinations of capacity options were initially generated) and the short amount of time available during the one-day workshop.

With so much work being carried out outside of the workshops, one might expect a lack of 'buy-in' from the workshop team. This would be to ignore the fact that many of the workshop team members have also been involved in these 'outside activities' (e.g.the generation of workshop information) and so may have a personal stake in the success of the workshop.

The quality of the Strategy information - thoroughly researched and accessible to all participants - may well have been an important factor in the success of the workshops. It meant that participants did not have to make do with estimates and local knowledge, as may otherwise have been the case.

D1.4. DISCUSSION

D1.4.1 Inclusion of Sustainability within the Strategy

Assessing Relative Sustainability

The relative sustainability of asset management options may be assessed by comparing their 'performance' against a range of sustainability indicators. The SWARD guidebook has categorised these sustainability indicators as technical, economic, environmental or social.

Although this type of formal sustainability assessment was not undertaken as part of the Strategy decision making process, it is clear that the Authority's assessment of options did consider - in different ways - the technical, economic, environmental and social impacts of the alternative options. The preferred package of Strategy options were selected according to a technical criterion (achievement of performance hurdles) and an economic criterion (whole life cost). However, it is clear that the Strategy is centred on an awareness of customer needs and the level of service that must be provided. An improved ability to provide water at a specified pressure and quality, even in times of drought, to sensitive customers (e.g.hospitals and industry) must represent an improvement in social sustainability. Likewise, the environmental sustainability of options was, to some extent, assessed through the identification of likely environmental constraints.

Although it appears that Strategy options were largely selected on technical and economic grounds, it is clear that social and environmental impacts were also considered. It is probably true that relative sustainability was, to some extent, implicitly assessed as part of the Strategy development process.

"Doing things right or doing the right thing?"

The 1996 Augmentation Study concluded that the only solution to the potential water deficit was to import water resources from an adjacent area, at a high financial cost.

From the outset, the Strategy questioned whether this really was the "right thing" to do. It concluded that it was not, and a range of other measures should instead be used to reduce the estimated 77 Ml/d water deficit (based on a 1 in 50 year yield). These can be summed up as "getting more for less" by improving existing assets and operational practices in preference to developing new assets. This resource efficiency is entirely in line with improved sustainability.

Leakage reduction activities alone will reduce the deficit by at least 35 Ml/d. Upgrading of water treatment works, boreholes and pumps; reinforcement of key network mains; and the development of new boreholes will allow resources to be used conjunctively, reducing the deficit by a further 29 Ml/d. Maximising an existing water import facility will reduce the deficit by another 4 Ml/d. The remaining deficit of 9Ml/d may come from supplying a major industrial consumer with raw water from an under-used Loch (subject to a feasibility study) and/or further leakage reduction arising from ongoing annual mains rehabilitation.

This package of options is undoubtedly more sustainable than the original water augmentation scheme.

Stakeholder Involvement

The SWARD Guidebook defines a stakeholder as "any individual or group which has a stake in how water and wastewater services are delivered, and could include customers, employees and community groups". In recent years, the term stakeholder has become less synonymous with shareholder and has been broadened to include those with social and environmental interests. It is also helpful to distinguish between internal stakeholders (employees of the Authority, including contractors) and external stakeholders (such as customers, environmental groups, community groups, etc).

It is widely accepted that decisions taken with the involvement of relevant stakeholders are more likely to be successful, in terms of quality and acceptance of the decision, than those taken without such involvement.

In the Authority's corporate plan states that it will work with (its) customers and a wide range of partners – (its) regulators, unitary authorities, statutory bodies and the community – to promote Agenda 21. (Agenda 21 being the international agreement, resulting from the 1992 Earth Summit, to take action to achieve sustainable development).

External stakeholders have been involved as consultees in the development of the Strategy. One of the Strategy documents, for instance, contains details of an initial consultation exercise conducted with a variety of statutory and non-statutory organisations. Included in these are: Scottish Natural Heritage (SNH), Scottish Environmental Protection Agency (SEPA), Scottish Wildlife Trust (SWT) and the Royal Society for the Protection of Birds (RSPB).

External stakeholders did not participate in the VM workshops. Of the workshop participants, approximately three-quarters were Authority employees and the rest were consultants employed by the Authority. There did, however, appear to be a spread of expertise within the Authority employees, suggesting that they were drawn from various functions of the organisation.

On the face of it, the non-participation of external stakeholders in the VM decision-making process is at odds with the principles of sustainable development. There are, however, plausible reasons as to why external stakeholders should have been excluded from this process.

Firstly, the workshops may have been limited to internal stakeholders for reasons of commercial confidentiality.

Secondly, the Strategy involved asking some fundamental questions about target levels of service, current and future yields, demand growth, etc. It is conceivable that some 'outsiders' could perceive this 'healthy internal debate' as indecision or even incompetence. Were external stakeholders to be involved at these early stages, the Authority might have felt compelled to present a united front, thus stifling genuine discussion.

Thirdly, many external stakeholders could reasonably ask of the Strategy "what has this got to do with me?" ie the high-level decisions taken as part of the Strategy may be of limited relevance to stakeholders until they filter down to become real, specific, local projects.

Future Developments
The following developments may, in future, improve the inclusion of sustainability in the Authority's decision-making processes.

Demand Management
The Authority has recognised the need for demand management, not only to increase water savings and reduce capital expenditure, but also to further protect the environment (for example by reducing unnecessary abstractions).

One of the Strategy documents discusses methods such as:

- installation of water Hippos to reduce the volume of water used in toilet flushing;
- rainwater recycling for garden watering and car washing, using water butts;
- grey water systems to recycle bath or shower water;
- domestic customer metering;
- education of all customers in the benefits of demand management.

Such methods are generally regarded as steps towards greater sustainability. Consequently, including them in a list of potential asset management options will improve the chance of a more sustainable solution being adopted.

The Authority is currently performing demand management trials within the Strategy area to assess likely water savings. The results from this study are expected in about a year.

Business objectives and key performance indicators (KPIs)
The Authority has investigated the use of a 'standardised' list of business objectives and associated KPIs (though not as part of the case study water strategy). The Authority's previous attempts at defining clear business objectives were analysed, taking into account the opinions of the Water Industry Commissioner (WIC), SEPA and the Scottish Executive.

The outcome of this analysis was a proposed list of business objectives that were categorised as either *customer-oriented* or *community-oriented* (as shown **in Table D1.4**). Although these objectives and KPIs have not been formally adopted, they do signify the direction in which the Authority was moving (WI Scotland, 2002).

Table D1.4: Types of business objectives

Customer-Oriented Objectives	Community-Oriented Objectives
Customer Serviceability	Environmental Sustainability
Operational Performance	Social Welfare
Asset Stewardship	Public/Political Acceptability
Financial Performance	
Revenue Performance	

Clearly, this spread of objectives takes into account the four categories of sustainability criteria proposed by the SWARD project : Economic, Technical, Social and Environmental.

Most of the objectives are further broken down into 'sub-objectives', which have associated KPIs. This is illustrated (for the customer serviceability objective) in **Figure D1.6**.

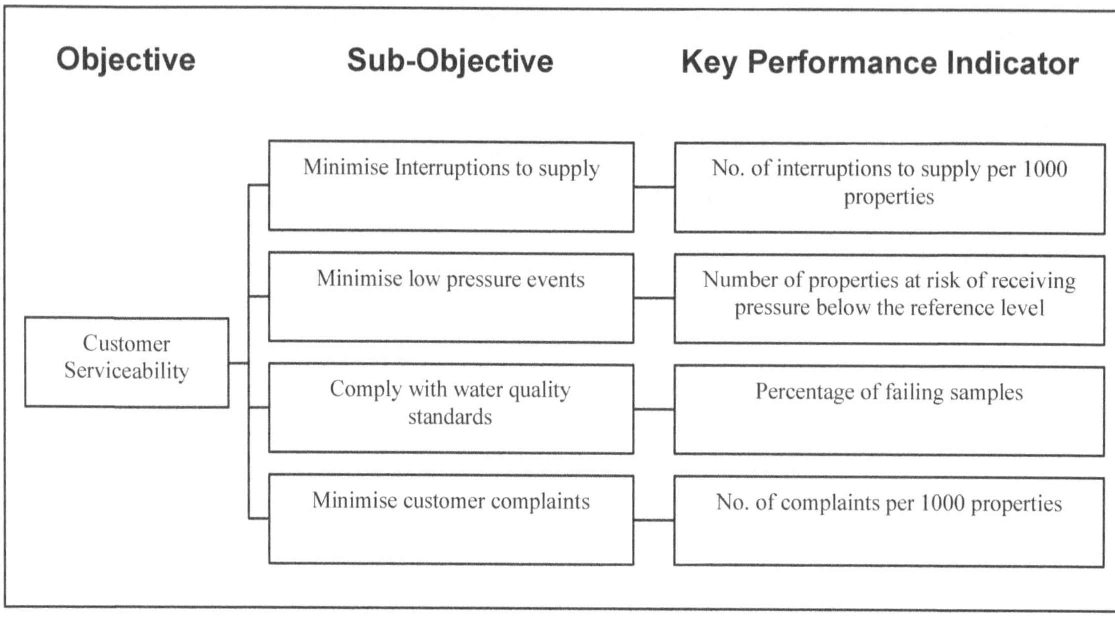

Figure D1.6: Hierarchy of business objectives

This is similar to the structure proposed in the SWARD Guidebook, i.e.

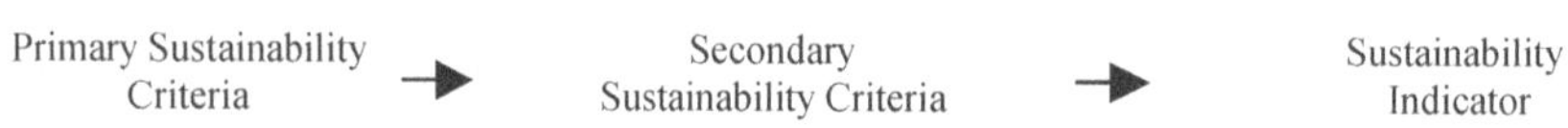

Alternative decision-making process

A draft decision making process was developed that is based on the business objectives and KPIs described earlier. This would be used to select a preferred combination of schemes.

It may be thought of as a type of multi-criteria analysis, using a spreadsheet model, the main features of which are outlined below.

Set Targets and weight KPIs:

- KPI values of a particular system set for the current year;
- KPI target values set for future years;
- expected KPI values set for future years assuming a 'do nothing' scenario;
- customer focused KPIs weighted by groups using Value Management;
- environmental/social KPIs weighted by specialists.

Enter impacts of alternative options (schemes).

- Spreadsheet based template used to collect cost/benefit information such as:
- financial costs & benefits;
- reduction in risk;
- quantitative impact on customer-focused KPIs;
- qualitative impact on environmental and social KPIs;
- confidence range for costs, benefits and KPI impacts.

A selection spreadsheet is then populated with the following:

- 5, 20 & 40 year Net Present Costs;
- Net Present Value of 40 year impact on each customer KPI;
- environmental/social scores.

Selection spreadsheet used to:

- calculate, for each KPI, the % of target achieved by each scheme and the % of budget taken by each scheme;
- experiment with various combinations of schemes;
- check the net present cost against available budget;
- check, for each KPI, the % of target achieved by the selected combination.

Environmental/Social scores may be used to choose between schemes with similar costs and which meet the customer-oriented KPIs. Similarly, the confidence range in the expected benefits and impact on KPIs may also discriminate between schemes with similar costs and impacts.

This decision making process is interesting in that it considers quantitative impacts (technical and economic), qualitative impacts (social and environmental) and uses a mixture of techniques such as scenario planning, cost-benefit analysis and value management.

The Authority has not yet adopted this process.

D1.4.2. Comparison: Water Strategy and SWARD Guidebook

The SWARD Guidebook decision making process is intended to "complement rather than replace existing decision making structures within the water industry". In order to discover how the Guidebook process might complement the case study Strategy development process, it is necessary to compare the two.

Purpose

To begin with, the stated purpose of each process is different. While the Guidebook centres on the explicit assessment of sustainability, the Strategy is more concerned with finding ways of enabling the Authority to "meet current and future demands" and "respond effectively to competition". These options are primarily assessed on performance and cost grounds, with less emphasis on social and environmental impacts. Any assessment of sustainability is therefore implicit – or, to put it another way, constraint driven rather than objective driven.

Phases

The main phases of the Guidebook are shown in **Figure D1.7**.

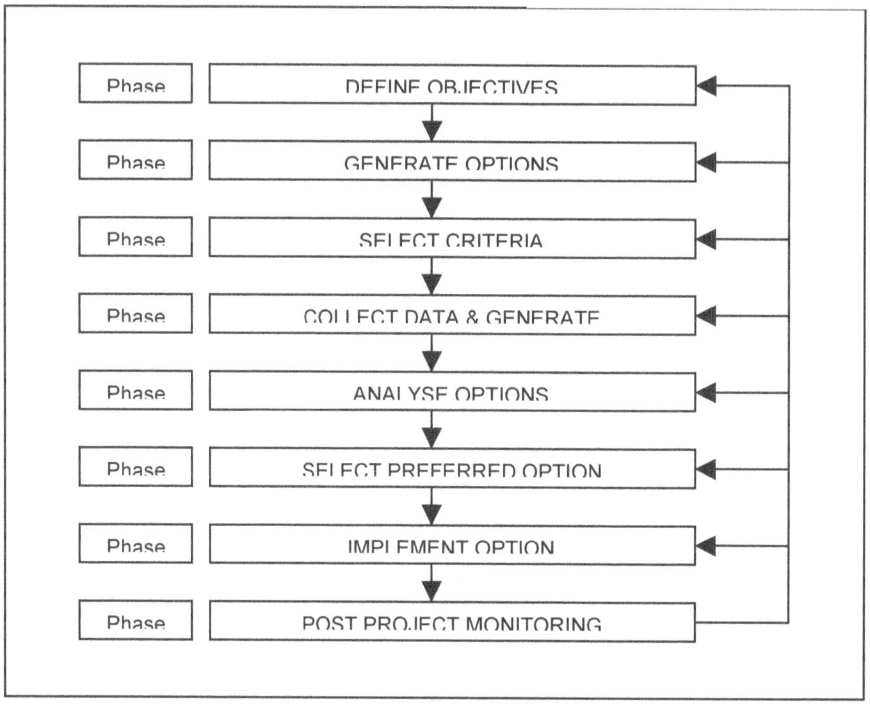

Figure D1.7: SWARD Guidebook Phases

These phases are essentially sequential, with the only feedback occurring after the post-project monitoring phase.

The Guidebook phases may be 'mapped' onto the phases of the Strategy Development Process, as shown in **Figure D1.7**.

From **Figure D1.8** it is clear that the Strategy Development Process can be represented using the phases of the SWARD Guidebook. It is also clear that the ordering of these phases within the Strategy is more complex than is suggested in the Guidebook. This can be illustrated by way of the following examples:

The Guidebook phase "data collection & information generation" actually occurs not once but three times in the Strategy process described in this Case Study, each time as a result of a meeting or VM workshop.

The "data collection & information generation" and "generate options" phases are carried out in parallel in the Strategy process (not sequentially as in the Guidebook).

Following VM2b, there is still a need to collect data and generate information (e.g.the production of control rules for conjunctive use). Thus, in the Strategy, a "data collection" phase can even occur after the "select preferred option" phase. This is not the case in the 'ideal' Guidebook-type process.

Additionally, there are elements of the Strategy process (not included in Figure 8) that do not fit easily into the Guidebook process. Such elements might include "identify and discuss critical issues" or "gain shared understanding of problem", both of which are typical VM workshop components.

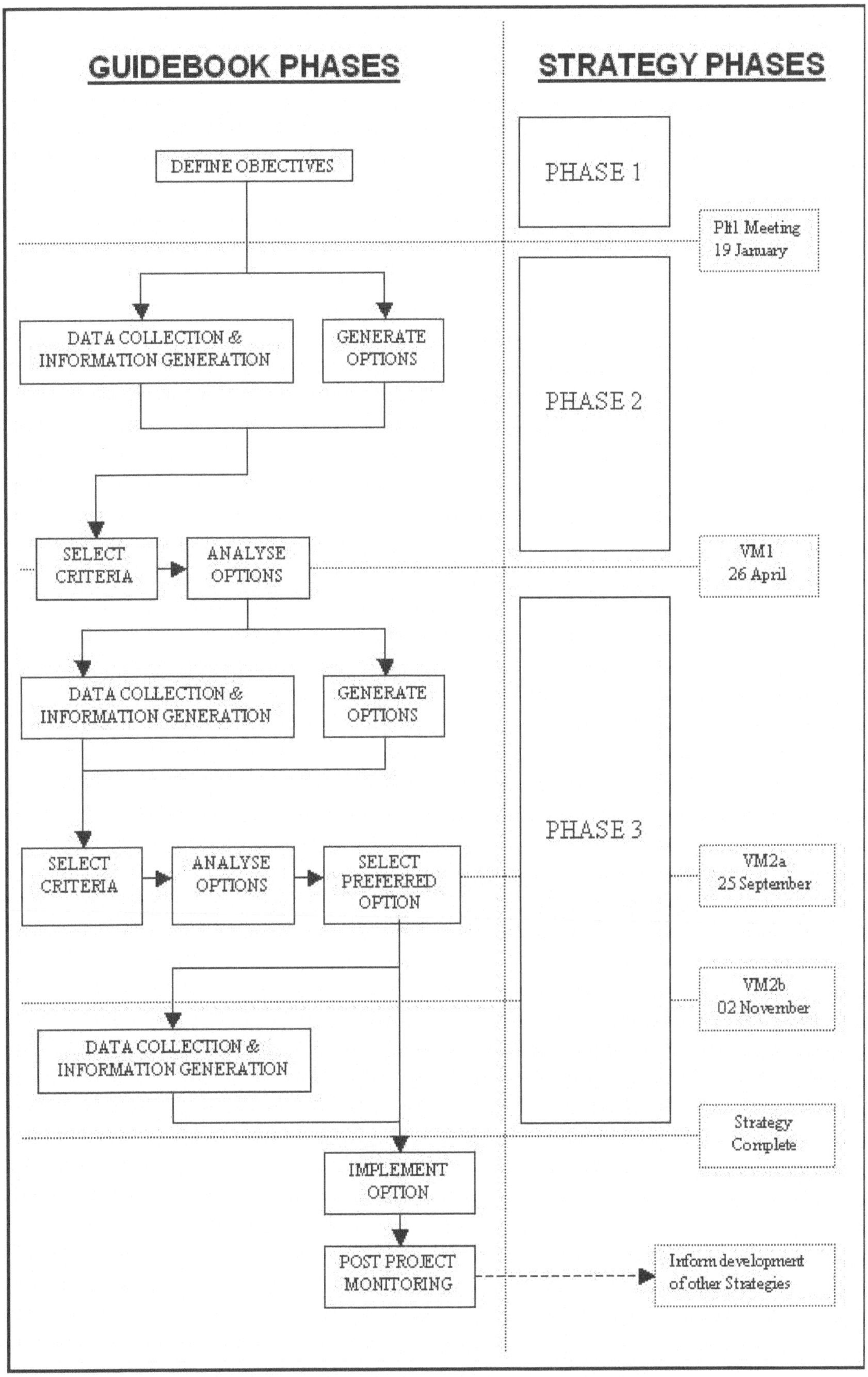

Figure D1.8: Guidebook and Strategy Process Phases

D1.5. SUMMARY AND CONCLUSIONS

Although the SWARD Guidebook might give the impression that the decision process consists of a series of phases undertaken in strict order and completed in one session (or at least, in a once through sequence), this is unlikely to be the case in real life except for the very simplest of problems. For instance, this Strategy Case Study took place over a full year and involved 3 VM workshops. The interactions between the many stakeholders were complex, involving a huge list of tasks that unavoidably took time to complete (e.g. waiting for software, running models, meeting with consultees, waiting for responses, talking to operations people, writing reports, preparing presentations, etc). Inevitably, externalities during this period sometimes led to changes of the base assumptions and the reordering of preferences.

It is therefore important to stress that the SWARD Guidebook Process is interpreted in that sense – as a guide. The context of a particular application, and the other exigencies that inevitably occur within a WSP will inevitably mean that the decision process has to be adapted to meet local and temporal circumstances and the influence of other ongoing developments with the business of the WSP. The key point is that most of the steps that make up the SWARD Guidebook Phases also occurred in the Strategy, though not necessarily in the same order.

Within the Strategy Process, VM was used main Procedural Tool and was the equivalent of the Guidebook Process. Within this, a number of Analytical Tools were used, most of which were hydraulic and associated modelling techniques, though formal MCDM tools were not used to determine the chosen Strategy. For this particular problem – the determination of a high-level water resources and distribution strategy – the VM approach was very effective and was broadly consistent with the SWARD Process.

REFERENCES

Bouchart, F.J-C., Foxon, T., Souter, N., Ashley, R., Butler, D. and Jowitt, P. W. (1999), Decision mapping for the development of multi-criteria analysis tools to assess the relative sustainability of water/wastewater systems, CHAINET Workshop, Seville, 26-28 March 1999 (available at http://www.sward.org.uk)

Bouchart, F.J-C., Blackwood, D.J. and Jowitt, P.W. (2002), Decision Mapping: Understanding Decision Making Processes; Civil Engineering and Environmental Systems Volume 19 Number 1, March 2002 Taylor and Francis, ISSN 1028-6608.

It might also be noted that the Water Industry (Scotland) Bill 2002, which established Scottish Water, places a statutory duty on the Authority in terms of sustainability: *"46B Sustainable development (1) Scottish Water must, in exercising its functions, act in the way best calculated to contribute to the achievement of sustainable development..."* HMSO 2002 Water Industry (Scotland) Act 2002 asp. 3

Appendix D2: Management of Domestic Sanitary Waste disposed of via the WC

Contents

D2.1 INTRODUCTION

D2.1.1 Overview of the case study

This case study illustrates the use of the decision support processes outlined in the SWARD Guidebook. It has been designed to demonstrate how sustainability can be incorporated into a theoretical asset management decision, and provides a fully-worked example of how the framework can be applied. The theoretical case study deals with the escape from the sewer system of domestic sanitary waste (SW) in a real catchment.

A number of possible options for dealing with the problem of SW escape and a variety of tools that may be used to inform preferred solutions are presented. The relevant sustainability criteria have been determined, and both qualitative and quantitative data assembled for each option, allowing determination of the 'most sustainable' of the options considered for SW management. The study utilises a number of formalised MCA approaches to the preferred option selection using computational models.

D2.1.2 Case study selection

The disposal of domestic sanitary solids using the WC as a 'rubbish bin' is habitual in the UK and stems from the historical link associating health risks with human waste (hygiene). Recent concerns about excessive water use for WC flushing, and the downstream costs of dealing with these solids have forced a re-appraisal to be made of this method of disposal. Social survey data and information from the Absorbent Hygiene Product Manufacturers Association (AHPMA) estimates that some 2.5 million tampons, 1.4 million sanitary towels and 700,000 panty liners are flushed every day. Of great concern is the increase in the amount of plastic being used in these items and changes in public usage patterns. Numbers and weights being disposed of are unlikely to reduce in the foreseeable future, even with government and other minimisation initiatives. The problems for wastewater system operators caused through disposal of SW items via the WC include:

- blockages;
- an increase in the requirement for sewer maintenance;
- problems for sludge disposal due to the plastic component of these solids, particularly where treatment works have communitors or macerating units;
- an increase in loads to wastewater treatment works (necessitating screens); and,
- significant impacts on the environment via overflow discharges and 'escape' through screens (see Figure 1) (Ashley et al., 1999).

There are several initiatives in the UK to encourage the disposal of sanitary refuse away from the sewered route into the solid waste stream. Welsh Water with Friends of the Earth has investigated the discharge to sewers of plastic objects and other debris (1993). A collaborative initiative between a wide variety of organisations including Water Companies, the then National Rivers Authority, manufacturers of sanitary products and environmental groups in the UK launched the national 'Bag-it-and-Bin-it' campaign in March 1995 [(1)] which is still operating in England & Wales. This is attempting to encourage the disposal of sanitary refuse away from the sewered route into the solid waste stream, using the campaign slogan of 'Bag It and Bin It, Don't Flush It'.

---

[1] Bag it and Bin it Campaign. Launched as a national initiative in 1995 by the National Bag It & Bin It Group. Members now include the U.K. Environment Agency, Tidy Britain Group, Women's Environmental Network, Anglian Water, Northumbrian Water, North West Water, and the Department of the Environment. The aim of the Campaign is to solve the problem of protecting beaches and rivers from sanitary waste by spreading the message of a simple solution: "Bag It and Bin It - Don't Flush It".

It is apparent that the best way of managing this waste needs to be determined both for the present and the future. A multi-disciplinary scientific study, funded by EPSRC, the Scottish Water Authorities and others, has previously been carried out to investigate the relative sustainability of the disposal of sanitary solids via either the WC or the solid waste routes (Ashley *et al.*, 1999). The original study considered the concept of sustainability for this issue and addressed the three main component areas: economy, environment and society, along with the detailed technical performance of alternative systems (flushing and binning) used for the disposal of SW (Swain and Souter, 2001). Large amounts of data were collected, and key management options investigated. Cost Benefit Analysis (CBA) was conducted to measure the net social benefit of reducing SW disposal to the waterborne route. In the present case study, the earlier research project is extended to consider a wider range of management options for dealing with the problem of sanitary waste escape from the sewer system to the environment.

D2.1.3 Study area selection

The original SW research was carried out in four test areas in Scotland, encompassing a range of settlements, from a high density urban housing scheme to a small inland rural community. Following on from the SW study, one of the Scottish Water Authorities commissioned the Urban Water Technology Centre (UWTC) at the University of Abertay Dundee to investigate in detail the application of the research findings to new areas of Scotland, one of which was the town used in this case study. Qualitative and quantitative data have subsequently been collected for the town, and it was therefore considered to be the most appropriate study area for the SWARD case study.

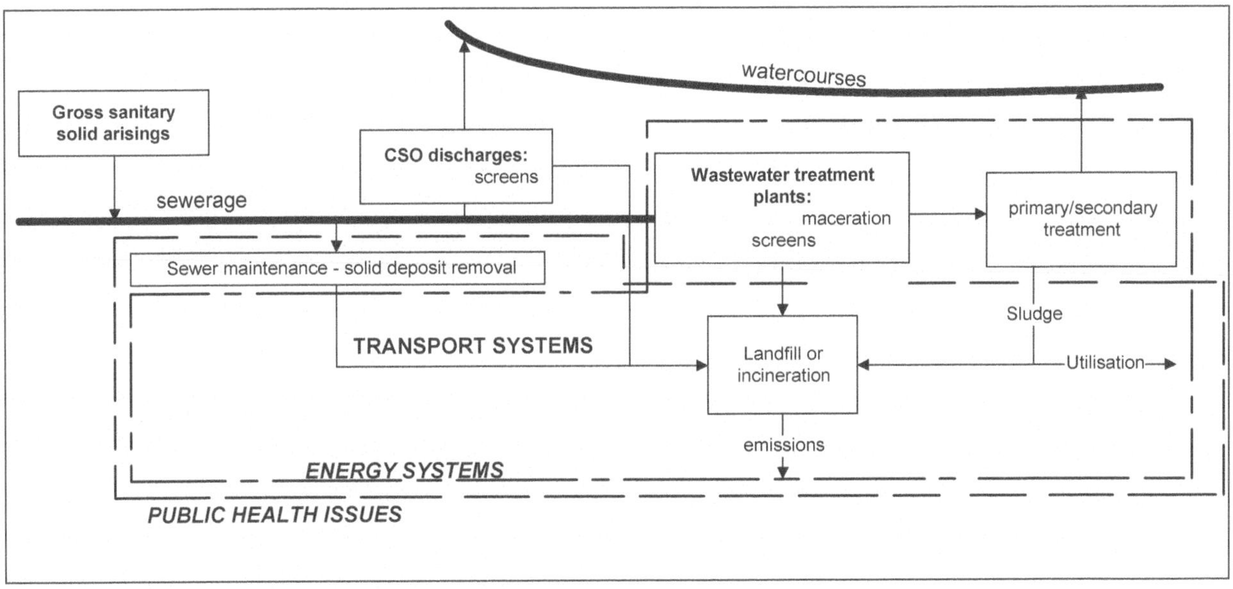

Figure D2.1: Gross sanitary solid disposal via wastewater stream

D2.1.4 Collaborator involvement

The case study has required active collaboration between the three Scottish Water Authorities and the SWARD case study team to produce a realistic exercise. The collaborators have been involved in each of the phases of the decision support process during the case study. **Table D2.1** outlines the range of collaborator involvement at these different stages, whilst **Table D2.2** details the type of data collected from a range of personnel. A steering group, made up of personnel from all three of the Scottish Water Authorities, held regular meetings throughout the duration of the case study and provided feedback and validation on all aspects of the work.

Table D2.1: Collaborator's role in the SW case study

PHASE OF DECISION SUPPORT PROCESS	COLLABORATOR'S ROLE
Definition of decision objective and determination of options/scenarios	-Feedback regarding the decision objective and the selection of options
Selection of pertinent criteria from generic criteria identified	-Feedback regarding criteria selection
Data collection	-Aid data collection -Provide data through consultation
Analysis	-Provide threshold values for each of the criteria identified (maximum and minimum acceptable) -Provide weightings for criteria selected -Provide preference scale and indifference scale

Table D2.2: Personnel interviewed for the case study

POSITION IN WATER AUTHORITY	AREA OF EXPERTISE	INFORMATION COLLECTED
Central Office Staff	Finance	Cost information e.g. capital costs, hourly rates for personnel, bills
Design Engineers	Wastewater Treatment Works	Hydraulic flow through works Purpose of each unit within works
Maintenance Supervisors	Repair and Maintenance of Equipment	Records of repairs and costs
Sewerage Network Managers	Sewerage Network	Impact of SW on sewerage network
Treatment Works Operatives	Maintenance tasks	Man hours to undertake individual maintenance tasks with specific reference to SW
Treatment Works Supervisors	Wastewater Treatment Works	SW flow through works Hydraulic flow through works
Wastewater Treatment Managers	Sewerage Network, Water Treatment Works	Overview of the wastewater systems

D2.2 THE STUDY AREA

D2.2.1 Catchment scoping

Site visits to the town allowed the options for managing SW discharges to be investigated in further detail. The first stage of the site visit involved:

- Establishing the natural drainage pattern in the catchment
- Establishing the type of surface (porous paving etc.)
- Establishing the housing types
- Type of residential drainage from roofs and drives
- Verification of sewer drawings

D2.2.2 Study area

The town studied is situated on the coast. The catchment has 626 domestic properties and a population of 1516 persons. It is a commuter town and most of the housing in the catchment is detached or semi-detached properties. These properties have large impermeable roof areas, and the majority also have drives of which about 30% are porous (paving or gravel).

A large portion of the town's developed area is on high ground above the original coastal floodplain. The steep gradient down to the Wastewater Treatment Works (WwTW) therefore has an effect on the drainage patterns of a large proportion of the impermeable catchment. Under storm conditions ponding is prevalent on flat impermeable areas. In sloping areas surface water flows freely, often avoiding road gullies and finding its own route to vegetation or to one of three streams in the catchment. The catchment is served by 80 % combined and 20 % surface water sewers.

D2.2.3 System definition

The 'system' is the sewerage system from the household WC to the town's Wastewater Treatment Works (WwTW) and its outfalls (**Figure D2.2**). The 'system' includes the stakeholders and sources of water from the water cycle, which interact with the sewerage system in the catchment. The system also includes the solid waste route in relation to SW disposal.

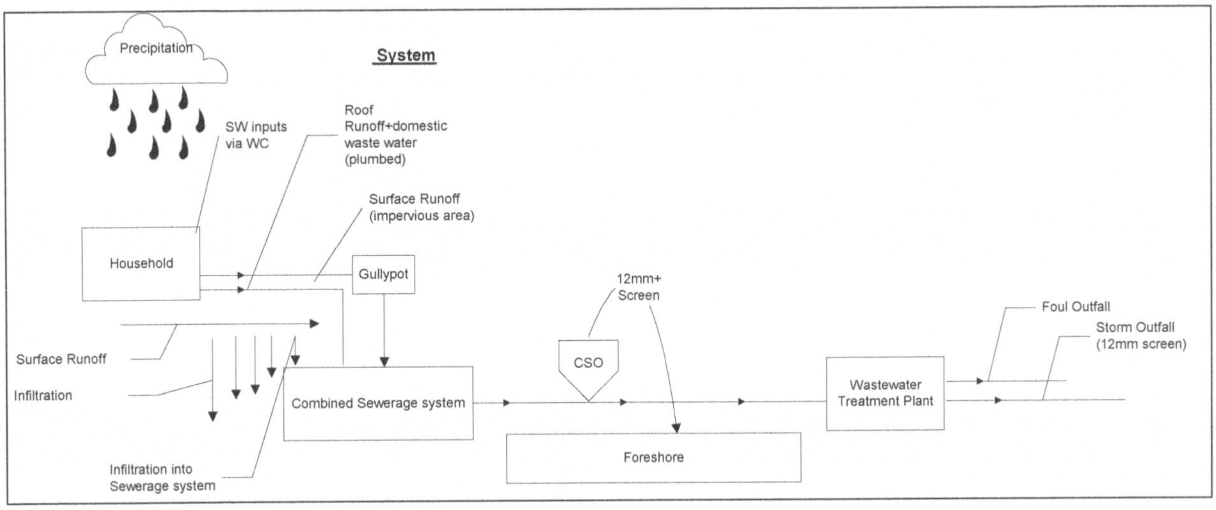

Figure D2.2: Diagrammatic illustration of the wastewater system for study catchment

D2.2.4 Sewerage system performance

Surface runoff from the hills directly behind the town contributes significantly to the amount of surface water in the sewer system. The infiltration from streams running through the catchment is also believed to contribute significantly to the flows in the sewerage system during dry and wet weather. It is estimated that infiltration increases flow to the WwTW by approximately 24 l/s. The network has an upstream CSO, which appears to be non-functioning; visual inspection showed no evidence of discharge. The CSO has also been found not to spill with a continuous one-year rainfall storm simulation using InfoWorks (1999). Other storm events were investigated e.g. 5 years, 10 years and 20 years and the CSO also remained non-functioning. This may be primarily because of the high weir level setting at the CSO. A storm outfall exists at the treatment works, with two storm pumps of 150l/sec discharge capacity each, plus an emergency outfall. A schematic diagram for the sewerage system is shown in **Figure D2.3**.

The schematic in Figure 4 shows the WwTW. It comprises of a pump house, screen house and septic tank complex. The pump house contains three foul pumps (1 standby) and three storm pumps (1 standby). The 2 foul pumps and 2 storm pumps have a combined maximum flow rate of 336 l/s. The screen house contains 2 drumscreens (1 standby), which screen up to 350 l/s. Screenings are removed, compacted and then disposed to landfill. The average weight of screenings for disposal is 11.2 tonnes p.a. Approximately 10% of the time the storm outfall is active and the flow is screened by 12 mm hand raked bar screens after being screened by the 6 mm drumscreens (see **Figure D2.4**). The foul and storm outfalls (the storm outfall also acts as an emergency overflow in case of pump failure) discharge to the ocean. Although the storm overflow is screened there is evidence of gross solids on the beach facing the treatment works and on the 12 mm hand raked screens.

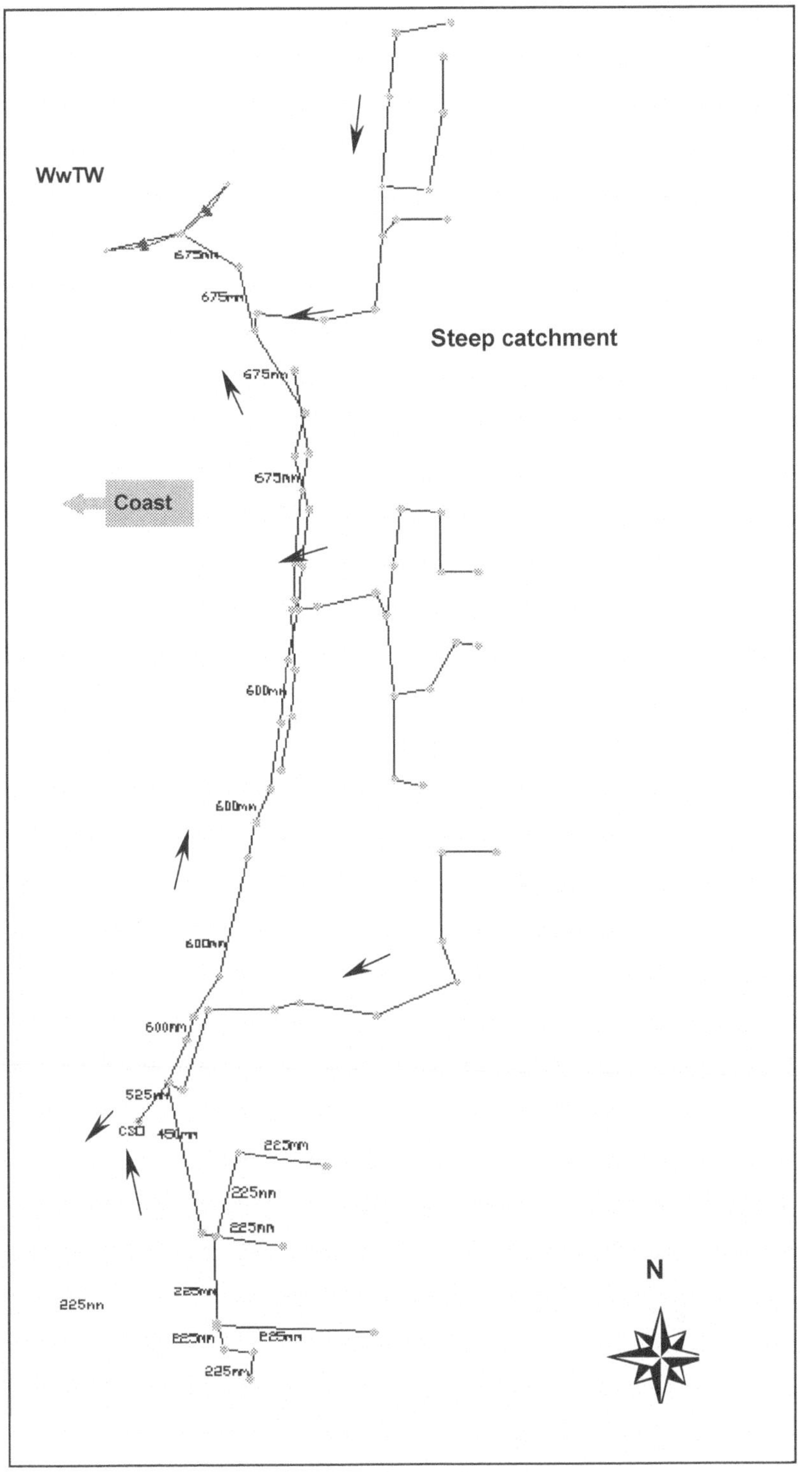

Figure D2.3: Sewerage system schematic diagram

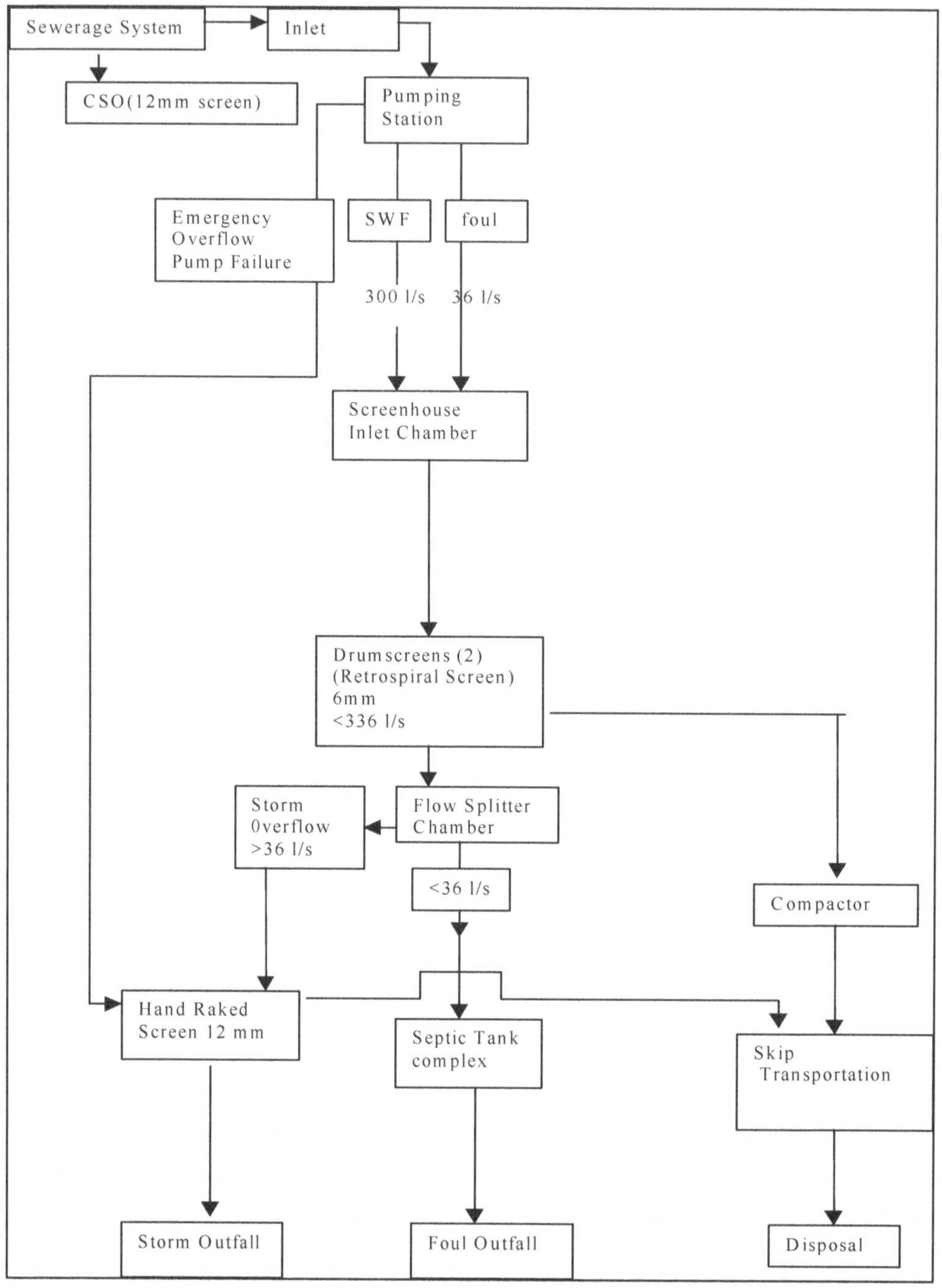

Figure D2.4: Wastewater Treatment Works Schematic

D2.3 SWARD PHASE 1: DECISION OBJECTIVE

The presence of SW in the environment will have a damaging effect on the amenity value of beaches in the town and the surrounding area. The decision for a problem relating to SW escape may be prompted by legislation or may be customer driven, i.e. a high number of customer complaints about the number of items of SW on beaches.

The decision objective in relation to the SWARD case study is:

> *'to reduce the number of sanitary waste items escaping to the environment via storm overflow discharges in the catchment in a way that is likely to be the most sustainable'.*

D2.4 SWARD PHASE 2: GENERATION OF OPTIONS

In Phase 2 of the DSP, SWARD team members acted as a design group to examine the decision objective and formulate a range of different approaches to the problem of SW escape. Six options for dealing with SW were generated by the group and these are presented in **Table D2.3**. The existing system can be modelled to incorporate each option individually to provide these different scenarios. The options 1-6 define the set of criteria that should be selected specifically for this case study. The options represent an attempt to encompass the widest possible range of approaches to tackling the problem. In an actual decision scenario, some of the options will probably have been deemed unfeasible by the decision-maker from the outset (as being too innovative, for example), and so it is likely that only two or three options would continue through to the ultimate multi-criteria analysis stages in reality. The RSS (reference sustainability system) (Foxon *et al.*, 2000) approach is used to examine the cost-effectiveness of using a combination of the six options generated here to tackle the problem of SW escape.

Table D2.3: The six options generated for the Sanitary Waste Case Study

OPTION	MEASURES	PRIMARY OBJECTIVE
1	Install 6 mm screens on storm overflow at WwTW	To screen storm overflows to constrain solids larger than 6 mm or equivalent to comply with minimum aesthetic pollution requirements for discharges
2	'Think Before You Flush' (TBYF) campaigns	To encourage a change in domestic disposal habits to dispose of sanitary solids in bin rather than WC
3	Install flow storage	To reduce frequency and volume of overflow spills thus reducing the number of SW items discharged to the environment
4	Retrofit stormwater source control	To reduce stormwater entry to the sewer system, thus reducing frequency and volume of overflow spills
5	Sewer rehabilitation	To limit infiltration to sewer system, thus reducing overflow spill frequency and volume
6	Retrofit outlet chokes on existing WCs and introduce these to new developments	To force a change in disposal habits from disposal via WC to bin due to the increased possibility of WC blockage in the home

D2.4.1 Generic approach to the problem

The six options address SW discharge at different points in the system. This is designed to allow a combination of the alternatives to be used if necessary. The six options approach the SW problem in three generic ways A, B and C, outlined below.

Approach A: End of pipe solution
This presumes that all solids entering the sewer network should be handled by the wastewater system. This is the traditional approach to wastewater management.

Option 1: Installing 6 mm screen (or equivalent) at WwTW storm overflow
The measures implemented as part of Option 1 were originally intended to include installing a new screen at the one CSO present on the sewerage system. This CSO was originally designed to discharge to the foreshore in the town (see **Figure D2.5**). This CSO was reported to have a 6 mm screen installed, therefore meeting the desired standard for aesthetic pollutants. During a site visit, the CSO was investigated and there was no evidence of recent discharge. The heights of the weir and screen diameter details were verified. Upon investigation the 6 mm screen that was believed to be installed was found to be of a larger bar spacing (12 mm) (see **Figure D2.6**). As there was no evidence of overflows at the CSO this was believed not to be of great significance, and replacement with a 6 mm screen was thought to be unnecessary.

The second overflow, at the WwTW, serves the storm/emergency overflow and is discharged to a long sea outfall. The storm overflow at the treatment works operates for 10% of the total discharge volume per annum, and is served by a 12 mm bar screen (**Figure D2.7**). Prior to reaching the storm overflow, flow should have been screened by the 6 mm screens installed in the main screenhouse, although there is some evidence of gross solids on the 12 mm screens.

As a result of the screening already conducted at the WwTW, this option actually represents the existing situation in the town. The objective of installing screens in this way is to meet the minimum discharge requirement for aesthetic pollution, which at present requires 6 mm screens or equivalent. The requirement for screening at overflows in the UK is a direct result of the almost ubiquitous disposal of gross solids via WC, and the subsequent aesthetic pollution of watercourses. In other European countries screening is unnecessary due to differences in disposal habits. This option presumes that all SW is conveyed to the WwTW and therefore in-sewer deposition has to be accounted for. The option only meets the present statutory requirement and the likelihood of the standard changing should be taken into account. Technical performance criteria will be selected to account for this aspect of uncertainty and changing standards.

Figure D2.5: The foreshore CSO

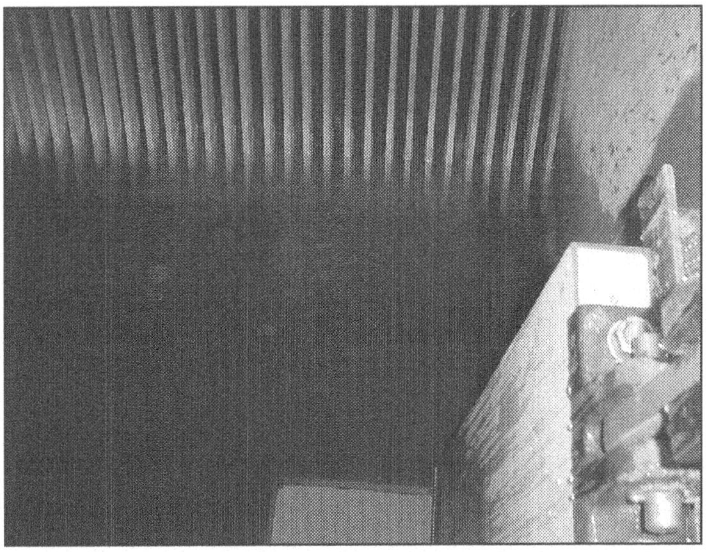

Figure D2.6: Screen installed in the foreshore CSO

Figure D2.7: Storm outfall chamber, Wastewater Treatment Works

Approach B: Habit change
Changes to inputs to wastewater systems may be effected by either or both: encouraging public behaviour to become more sustainable in use of these systems (WC, kitchen sink and other wastes, surface water discharges), or by re-engineering systems to exclude more of the inputs from WCs and surface water drainage.

Option 2: Think before you flush (TBYF) educational campaign
The 'Think Before You Flush' campaign highlights the issue of SW and educates the public about the effect of flushing SW [(2)]. Encouraging alternative disposal practices should reduce the number of items disposed via the WC.

The objective of this option is to achieve a sustained reduction in the number of SW items disposed by the waterborne route thereby removing them from the waterborne system. A TBYF campaign has already be carried out in the town as part of a project for the water authority. The information associated with this option is therefore readily available.

Option 6: Forced approach to habit change.
Retrofit/fit small-bore outlets to existing/new WCs
Small-bore outlets on WCs stop the flushing of large items of SW, forcing the public into using alternative methods of disposal. Large items of SW are physically prevented from entering the system, and there may also be some education of the public regarding the undesirable habit of flushing smaller items e.g. cotton buds. The objective of this option is to achieve a sustained reduction in the number of SW items disposed by the waterborne route, removing them from the waterborne system before they even enter. However, public acceptance of such technologies is required for them to be successful.

Similar initiatives have been carried out elsewhere, with one of the most ambitious schemes launched by the New York City Department of Environmental Protection (DEP) in 1994. In the first 25 months of the program, 1.3 million toilets were replaced at a cost of $290 million. The City of New York has now reduced the flow of water through its sewerage treatment plans by 80 million gallons a day or 2.8 billion gallons a year and saved the City $3 billion in water and wastewater treatment expansion costs. The average household in NYC is now saving $70 annually on its water and sewer bills, which are combined in the City of New York [(2)].

Approach C: Spill reduction

These options are designed to reduce flows in the sewer system by excluding some of the surface or groundwater. The overflows will then operate less frequently, discharging less SW to the environment.

Option 3: Install Sewer Storage
Increasing storage within the sewerage system will reduce the effect of increases in flow due to rainfall. Storage provides flow equalisation and reduces the peak flow rate. Storage may be in the form of on- or off-line storage tanks. Flow stored in the tanks is then drained and treated at the WwTW as the storm recedes.

A suitable site for the location of storage was determined during a site visit to the town. The town is a very dense development and therefore space to install storage was limited. The most suitable site was immediately south east of the WwTW. At the present time this area is a car park and viewing area, with an access road to the treatment works. Disruption and inconvenience to the public would therefore be kept at a minimum. Online storage was expected to be the most suitable option for the system, with a hydrobrake to regulate flow to the WwTW.

The objective of this option is to reduce SW carrying storm overflow spills, thus reducing the number of SW items in the environment.

Option 4: Source control.
Sustainable Urban Drainage Systems (SUDS) can be used to reduce stormwater input to the combined sewer. Such source control systems are typically situated immediately alongside the surfaces they serve. Several types of source control structures exist and these are generally used in combination (CIRIA, 2000). An example of some of the types of source control that could be used to retrofit the town are porous surface/car parks, roadside swales, attenuation control structures, and minimising directly connected impervious area (MDCIA) in the form of rainwater barrels or roadside gravel/French drains.

---

[2] For more information on the retrofitting of Ultra Low Flush Toilets in New York see:
http://www.voltviewtech.com/NYCDEP.HTM or http://www.epa.gov/owm/toilets.htm

The objective of the SUDS option is to reduce storm overflow spills by reducing the surface runoff of the catchment during a storm event. This flow attenuation will reduce the number of overflow discharges thus reducing the number of SW items discharged.

The large amount of porous paving and gravel driveways already installed in the town limits the scope of the changes that can be made. The infiltration capacity of the catchment (WRAP) is not very suitable for infiltration devices, and the density of the housing means the potential for retrofitting of swales is limited. The gradient of many of the roads prohibits the removal of curbs to allow natural runoff due to erosion. Roadside gravel drains or swales in some areas could possibility be viable but are not considered as part of this option. The large roof areas provide a good opportunity for minimising the directly connected impervious areas via rainwater barrels.

For some of the analyses outlined in the subsequent sections, a range of SUDS designs are used within this option. The proportion of houses for which rainwater barrels were a suitable option were estimated (i.e. those without barrels, with garden or yard areas). Calculations for Life Cycle Analysis (LCA) and for the mass balance of gross solids were done on 100% of these houses installed with barrels. However, it was considered that this would be an infeasible target in a real situation, and so calculations were also done for 35% and 28% of the total roof area controlled. For the final MCA the 35% sub-option was utilised.

Option 5: Sewer Rehabilitation

The infiltration/inflow into the sewerage system is believed to contribute 24 l/s to the flows during both dry and wet weather. Surface water infiltrating the sewer system increases the flow and therefore increases storm overflow spills. A concerted programme of sewer rehabilitation can achieve the reduction of infiltration/inflow. A site visit identified a number of streams in the catchment. Some of these are believed to contribute significantly to the flow. Surface water from unpaved areas entering road gullies is also a significant contributor to in-sewer flow in the catchment (see **Figure D2.8**).

Although a 100% reduction in infiltration is a laudable aim, it is unworkable in reality. A range of sub-options was also considered: 11%, 18%, 30%, 42% and 72% infiltration reduction. For the main MCA, the 42% reduction was judged to be a realistic target. This involved the rehabilitation of 78 m of sewer in the town, pinpointed by the water authority through dye testing, as contributing some of the greatest amounts of infiltration to the network. Rehabilitation in this case study refers to complete replacement of the old sewer, not patch repairs or other rehabilitation methods.

Figure D2.8: Wetland (to right) draining to road gully connected to the combined sewer, on the main road in the town

D2.5 SWARD PHASE 3: SELECTION OF PERTINENT GENERIC CRITERIA

D2.5.1 Procedure for selecting case study criteria from generic list

The case study utilises criteria that have been selected to define relative sustainability. Quantification of the attributes of the criteria where possible, allows comparisons to be made. Preferences may also be used for both quantitative and qualitative attributes. The case study criteria have been selected from (or informed by) the generic criteria to be relevant to the decision objective and to be applicable to all of the six options under consideration. In this case the objective was to reduce the number of SW items, therefore, all criteria selected must reflect changes in the system in relation to SW allowing the options to be compared. A criterion must apply to all of the options to be valid. Many of the criteria have been developed specifically for the SW case study.

The first step was to select all of the primary criteria that may be suitable for inclusion for the study. The secondary criteria that were related to the primary criteria selected, were also assessed. It is important at this stage to ensure that none of the criteria selected account for the same factor. Those criteria that were deemed not to be relevant were then eliminated. The criteria were selected to be independent. Having assessed the list of draft criteria, the next step was to identify any major factors that were not represented by the primary and secondary criteria list. In any decision-making situation, it is required to develop criteria specifically for the decision objective. An example of this in this case study lies within the technical 'Performance' category, where a criterion to measure the performance of the system to deal with sanitary waste was not present in the generic list. A criterion was therefore developed for this aspect of the systems performance, focusing only on sanitary waste escape. **Figure D2.9** illustrates the iteration process used in the selection of criteria for the case study.

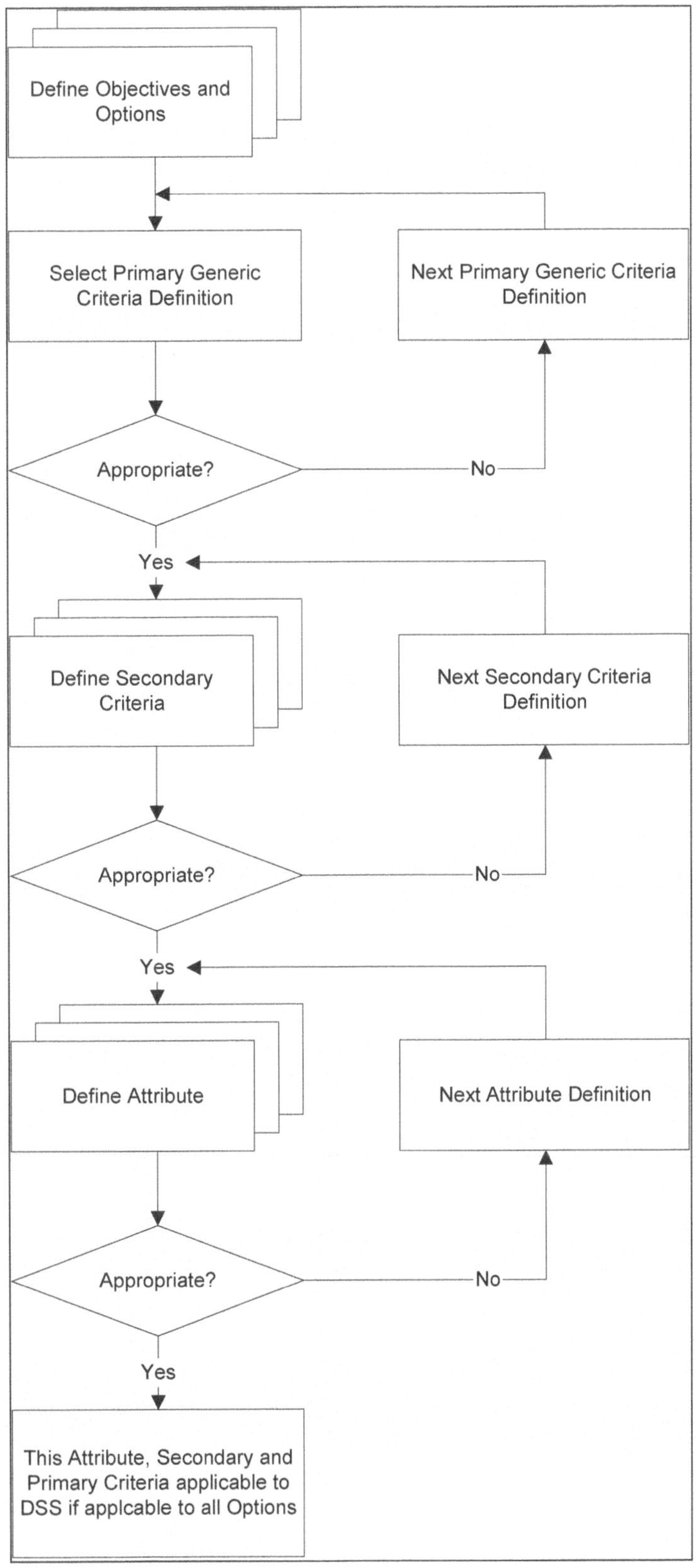

Figure D2.9: The iteration process used for criteria selection

D2.5.2 The criteria

Tables D2.4 – D2.7 present all of the criteria used for the SW case study, along with the method used to measure the indicator, and the units of measurement.

Table D2.4: SWARD Economic Criteria

PRIMARY SWARD CRITERIA	SECONDARY CRITERIA (& CRITERIA CODE)	INDICATOR	METHOD OF COLLECTION
LIFE CYCLE COSTS	E1: Capital cost: Investment	£ per catchment	Cost estimation for investment -consultation
	E2: Operation Costs	£ per catchment per annum	Cost estimation - consultation
	E3: Maintenance Costs	£ per catchment per annum	Cost estimation for investment -consultation
FINANCIAL RISK EXPOSURE	E4: Financial risk exposure	Qualitative expression of estimated risk	Consultation

Table D2.5: SWARD Environmental Criteria

PRIMARY SWARD CRITERIA	SECONDARY CRITERIA (& CRITERIA CODE)	INDICATOR	METHOD OF COLLECTION
RESOURCE UTILISATION	EV1: Energy Use	Total energy (MJ) for 20 year life cycle	Life Cycle Analysis
ENVIRONMENTAL IMPACT	EV2: Impact on air	Total CO_2 emissions (kg) for 20 year cycle Total NOx emissions (kg) for 20 year cycle Total SO_4 emissions (kg) for 20 year cycle	Life Cycle Analysis
	EV3: CSO/SO discharges to the environment	Volume per annum (cu.m)	InfoWorks modelling

Table D2.6: SWARD Social Criteria

PRIMARY SWARD CRITERIA	SECONDARY CRITERIA (& CRITERIA CODE)	INDICATOR	METHOD OF COLLECTION
ACCEPTABILITY TO STAKEHOLDERS	S1: Acceptability to stakeholders (customers)	Percentage acceptability (Qualitative)	Door-to door Questionnaire
	S2: Perceived impact on the environment	Percentage perceiving negative environmental impact (Qualitative)	Door-to door Questionnaire
PARTICIPATION AND RESPONSIBILITY	S3: Participation and responsibility	Level of participation required for option, High –low (Qualitative)	Consultation with water authority staff/SWARD team members

Table D2.7: SWARD Technical Criteria

PRIMARY SWARD CRITERIA	SECONDARY CRITERIA (& CRITERIA CODE)	INDICATOR	METHOD OF COLLECTION
PERFORMANCE	T1: Sanitary Waste Escape	Quantity discharged to receiving watercourse per annum (kg)	InfoWorks modelling, Mass Balance + screen performance information
	T2: Sanitary Waste Transport In Sewer	Amount of SW detained in sewer system per annum (High – low)	Consultation with water authority staff/SWARD team members
RELIABILITY	T3: Risk of Failure to provide service	High – low (Qualitative)	Consultation with water authority staff/SWARD team members
FLEXIBILITY AND ADAPTABILITY	T4: Flexibility and adaptability	Ability to accommodate future needs (Qualitative)	Consultation with water authority staff/SWARD team members

D2.5.3 Validation of criteria

The criteria that were selected by the SWARD team were discussed and validated at a steering group meeting with senior representatives of the three Scottish Water Authorities.

A workshop was also held involving eight members of staff from the water authority with the aim of weighting the importance of the selected criteria in the decision making process. The personnel involved were from a number of different departments and included the following:

- Capital and Tax Accountant
- Research and Development Advisor
- Contracts Operations Advisor (Networks)
- Environment Section (Advisor)
- Research Assistant (Research and Development)
- Team Leader – Operations (Wastewater Treatment)
- Investment Planner
- Contracts Operations Advisor (Wastewater Treatment)

A range of activities was completed in the workshop, involving individual ranking, placing the criteria into a category, and ranking the four categories of sustainability criteria. The final ranking of the case study criteria is shown in **Table D2.8**. This was obtained from the mean of all of the individual rankings. A group discussion took place at the end of the workshop in which the group rankings and categorisation of criteria were agreed. The inter-category importance determined by the workshop participants is shown in **Figure D2.10**.

In this way the SWARD team could determine the weighting of the criteria for both situations, namely single decision makers and group decision makers, thus allowing a comparison of the two to be made.

Table D2.8: Ranking of SWARD Generic Criteria

SWARD criteria	Overall Rank	
Capital costs	1	Most important
Risk of failure	1	
Acceptability to stakeholders	2	
Operational costs	3	
Maintenance costs	4	
Financial risk exposure	5	
Perceived impact on the environment	5	
Participation and responsibility	6	
Flexibility and adaptability	7	
Primary energy use	8	
CSO discharges to the environment	9	
Sanitary waste escape	9	
Sanitary waste transport in sewer	10	
Impact on air	11	Least important

D2.6 SWARD PHASE 4: DATA REQUIREMENTS AND COLLECTION

The previous detailed study of the catchment undertaken by the UWTC produced a large amount of data defining the performance of the wastewater system, determining the inputs of SW, and establishing public attitudes towards wastewater and sanitary waste disposal. The existing sewer system has also previously been modelled using InfoWorks. Data collection for the SWARD case study therefore involved collating this existing information, along with the collection and modelling of new data. The following section is divided up according to the four major categories of criteria, with the methodology used to determine values for each individual criterion discussed in turn. The complete set of results for all SWARD criteria are presented at the end of this appendix in **Table D2.25**.

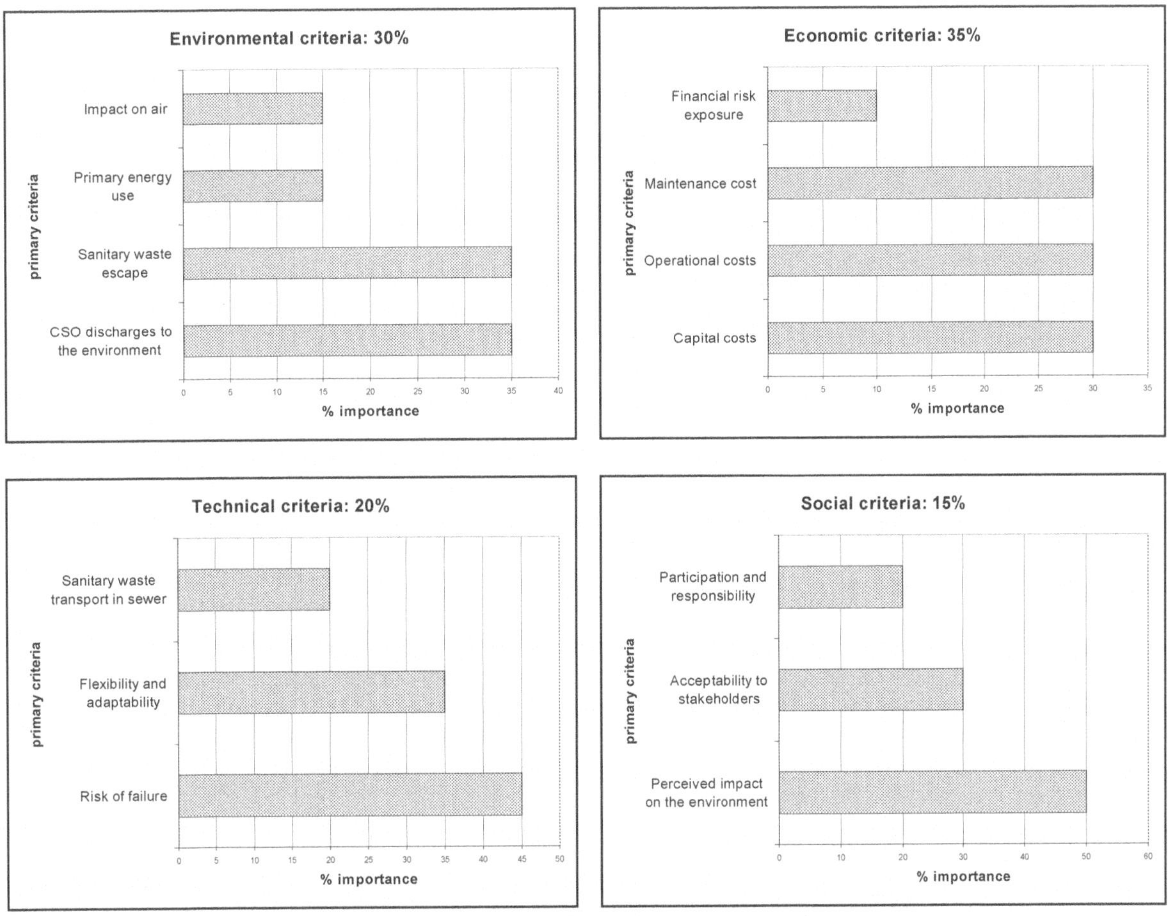

Figure D2.10: Collective prioritisation of the SWARD primary/generic criteria

D2.6.1 Economic criteria

Semi-structured interviews were conducted with a range of different levels of personnel from the water authority to gather data for the economic criteria. The interviews were held at the water authority premises and individuals were selected because of their area of expertise. Information was required on processes, maintenance protocol and plant design, with identification of the revenue costs directly attributable to SW in the waterborne system for the town. The presence of SW material in the waterborne system adversely affects the operation of certain components within the WwTW. Much of the indicator data consisted of measurements from meters installed at the treatment works. The water authority installed these meters for 2 weeks at the works in order to determine the amount of energy required for normal plant operation.

E1: Capital Costs

For each investment, capital costs were collected directly from the suppliers of the equipment and materials. The cost of installation were estimated and included in the final cost. Costs of contractors and externalities such as excavation were also included. The capital costs for each option were as detailed in **Table D2.9** below.

Table D2.9: Capital costs for the six major options

Option		Details
1	Install 6 mm screens on overflows at WwTW	The cost of a drum screen, including installation, was £68,000. The screen model was selected for the appropriate flow volume from the screen suppliers of the WwTW.
2	Think Before You Flush' (TBYF) campaigns	The campaign for the catchment population cost £10,000. This includes materials, transport and personnel costs.
3	Install flow storage	Sewer storage cost £70,000 for excavation, materials and construction.
4	Retrofit stormwater source control	157 rainwater barrels at £25 each are required for the target reduction. The total cost for the catchment for purchase and delivery is £3295.
5	Sewer Rehabilitation	This requires the replacement of 78 m of sewer at varying diameter and depth. The average cost is £500 per meter = £39 000 for excavation, materials and construction.
6	Retrofit outlet chokes on existing WCs and introduce these to new developments	This involves the retrofitting of 522 WCs in the catchment. Materials and installation costs are £43 per WC, with an additional £10,000 for the complementary educational campaign required.

E2 & E3: Operation, Repair and Maintenance Costs

The cost required for each option was established by estimating the likely maintenance and operation costs for each option. The operational and maintenance costs presented are those for the water authority ONLY. For example, option 4 (source control) is shown as having zero maintenance cost, despite the fact that there may be some occasional costs incurred for the householder. Respondents to the questionnaire administered for criteria S1 and S2 were made aware of which of the options included any maintenance (and hence financial) responsibilities on their part. The effect of any personal financial responsibility should therefore be included within the criterion 'Acceptability to stakeholders'.

- The operation and maintenance costs for the screen option are due to the presence of SW material in the waterborne system. Plant personnel provided detailed information with respect to the operation of the screen within the treatment works. They outlined the maintenance and repair problems commonly associated with the units and then identified which of these problems were as a direct result of SW. The accounted costs, were therefore those only directly due to the presence of SW. Maintenance costs comprised of man-hours relating to daily, weekly and annual tasks for each unit e.g. hose down the screen, screen cover, and remove screenings (2 men, 5 mins, daily). Repair costs were also included, and comprised of man-hours (e.g. hourly rate for mechanics) and costs of replacement components (e.g. replacement of bearings and seal on a drum screen).
- The operation and maintenance cost for the TBYF campaign consists of an annual refreshment campaign to reinforce the TBYF message. This costs £800 per annum.
- The maintenance costs for option 3 (install storage) represent the costs for jetting the storage tank twice per annum to remove any sediment and other deposition.
- There is no maintenance cost to the water utility for options 4 or 5.
- Option 6 has an annual cost of £800 for educational material and enquiry costs.

A summary of total costs for the six options is detailed in **Table D2.10** and also presented in Table 25 with the full suite of criteria values.

Table D2.10: Summarised economic costs of the six options for SW management

Option	Description	Capital £	Annual Operational (running) £	Annual repair and maintenance £
1	Install 6 mm screens on overflows at WwTW	68 000	628	3167
2	Think Before You Flush' (TBYF) campaigns	10 000	800	0
3	Install flow storage	70 000	0	680
4	Retrofit stormwater source control	3 295	0	0
5	Sewer Rehabilitation	39 000	0	0
6	Retrofit outlet chokes on existing WCs and introduce these to new developments	32 446	800	0

E4: Financial risk exposure

Senior management, acting as the decision-maker for the case study, were consulted in order to obtain a qualitative expression of the considered risk. In reality decision makers are frequently faced with scenarios where many of the data obtained consist of qualitative judgements. Many decision support methods then require the translation of these expressions into quantitative values. The use of this qualitative data as an input into the MCA model used in this case study is acceptable, as the software has been designed to respond and incorporate this type of information.

D2.6.2 Environmental Criteria

EV1 & EV2: Energy Use and Impact on air

The use of Life Cycle Assessment (LCA) is a key component of the case study, and an essential tool in assessing the first two environmental criteria for all of the options. LCA was undertaken using the SimaPro software (PRé consultants B.V., 1997) to generate information about likely environmental impacts. Life cycle studies can be met by the analysis of a full life cycle, partial life cycle, or of individual stages or processes. A full life cycle approach is used in this study, as LCA must look at the overall service provided by the product to the user. The options are assessed from "cradle to grave" to include the different energy requirements and future material demands for the different solutions. The life cycle of each option has been assessed individually identifying: the materials used; energy consumed; and releases to the environment, and encompasses transportation, recycling and re-use. Assumptions have had to be made in relation to the design life of a particular product and the likely replacement parts requirements.

Data sources and data quality

The standard SimaPro4 Database has been used in this study. An important feature of SimaPro is the analysis techniques that it incorporates. A description of the main analysis techniques: characterisation, normalisation, evaluation and indicators are given in **Table D2.11**.

Table D2.11: SimaPro characterisation, normalisation, evaluation and indicators

CHARACTERISATION	In this technique, impacts are sorted into classes according to the effect they have on the environment. The environmental impacts are aggregated within each class to produce an effect score by multiplying extractions and emissions by an equivalence factor (Virtanen *et al.*, 1995). It is not normally sufficient to just add up the quantities of substances involved without applying weightings, since some substances may have a more intense effect than others. This problem is dealt with by applying weighting factors to the different substances.
NORMALISATION	In this stage the effects are divided by a 'normal' effect, for instance the effect caused by an average person during a day or any other period. By doing this it is possible to see the relative contribution from the material production to each already existing effect. The graphs created show the normalised effects of an average European inhabitant over one year. The impacts are compared on a relative scale of inhabitant equivalents. However, normalisation only reveals which effects are large and which effects are small in relative terms. It says nothing of the importance of these effects.
EVALUATION	Normalisation improves insight into the results. However, no final judgement can be made because all effects are considered to be of equal importance, which is not representative of the true effect on the environment. In the evaluation phase the normalised effects scores are multiplied by weighting factor representing the relative importance of the effect. This is represented in graphical form by the height of the column for each alternative. The evaluation graph applies a weighting factor, scaling the results to a certain level of seriousness. This seriousness is measured in indicator points, and is derived from the assessment of damage inflicted upon human health and on ecosystems in the Eco-Indicator 95 report.
INDICATORS	The indicator evaluation technique produces a single figure for environmental impacts. SimaPro can add together all the different environmental effects from the evaluation stage to give a total impact for each material and process in the assembly. The value mode in SimaPro displays the effect scores in tabular form. This option allows four interpretations of the effect scores to be displayed. This enables the precise contribution of the effect score for each component of the 'box' or process to be analysed.

System Boundaries

The boundary condition refers to what is included and excluded from each step in the LCA process. In this study the materials, energy, natural resources, transportation, use and disposal are analysed. The main sewer network and treatment works (other than screens) have not been considered in the analysis as the infrastructure is the same for each of the options. Only new infrastructure associated with the control of the sanitary solids via the options is investigated.

The scope and purpose of the project defines the system boundaries that are set. The LCA is being undertaken to compare the options available for the WSP, therefore the outputs required for MCA are one of the main requirements. The options are compared in relation to a number of indicators from the SimaPro package. This allows an assessment of the relative impact (part of the relative sustainability) of the six alternatives. A unit function of total gross solids reduction during the life time for each option is used to compare between the emissions from the different options.

As the life cycle for each option is assessed independently from the combined sewerage system, it is not necessary to determine a boundary for the whole sewerage system. Two of the options reduce the number of items entering the sewer system (user behaviour), three options reduce the flow in the combined sewer system (and hence overflow spills), and one option is an end-of-pipe solution (designed to receive and handle all items put into the system). However, the boundaries that are set for the options must be identical if a comparison is to be made. For each option the boundary has been determined as given in **Table D2.12**.

Table D2.12: System boundaries employed for the case study

TEMPORAL	Each option has a different design life, but for standardisation purposes each option was assessed over 20 years.
SPATIAL	The data were collected for the areas in which the components of each of the options were manufactured, and for the catchment under study for when the option was in place. The SimaPro database LCA model has been used where data was not available or is still under investigation locally.
RAW MATERIALS ACQUISITION AND ENERGY	The boundaries for this element do not include all activities needed for the acquisition of all raw materials. Nor does it include the first manufacturing or processing stage that refines the raw materials. Detailed process data were unavailable for all the options investigated. The SimaPro database has been used to supplement the detailed process information.
MANUFACTURING, FORMULATION AND PROCESSING	The processing step takes the raw materials and converts them into products. Data from SimaPro has been used for this stage except where factory specific process information was available.
TRANSPORTATION AND DISTRIBUTION.	In this stage the product undergoes physical change in location or configuration. This does not involve a transformation of materials. Transportation includes the movement of materials and energy involved. The transportation and distribution stage also involves those activities used after manufacturing.
USE, REUSE AND MAINTENANCE	The boundary condition for this stage begins after the distribution of product or materials and ends at the point where the product or materials are discarded and enter a waste management system.
RECYCLING AND WASTE MANAGEMENT	The recycling step involves reclaiming materials out of the waste management system and returning them to the manufacturing or processing stage. Waste streams are generated at each phase of the life cycle. Waste management includes any mechanisms for treating, handling or transportation of waste prior to their release into the environment.

LCA data requirements and collection

The main data requirements for LCA are (see **Figure D2.11**):

- Material
- Energy
- Transport
- Processing
- Use
- Waste generated
- Waste treatment

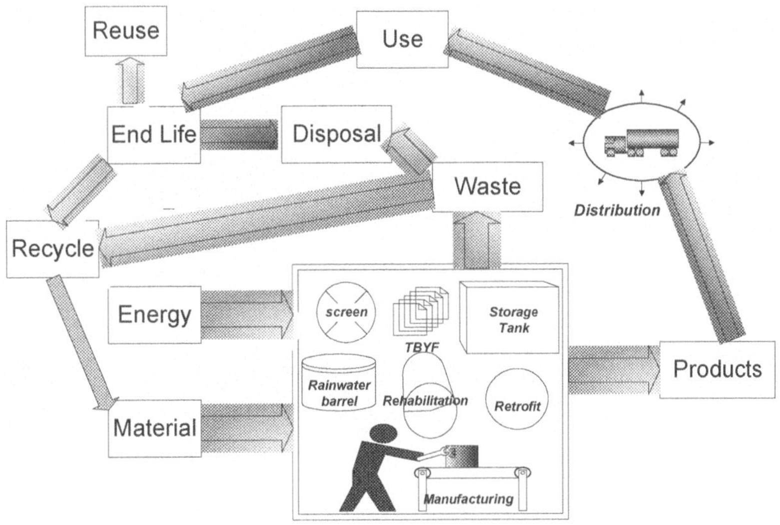

Figure D2.11: The elements of life cycle inventory analysis for the options

The data for the LCA have been collected from various sources. The manufacturer of the products used for each option were identified and contacted. These communications involved phone, fax and email contact, as well as site visits. The processes involved in the manufacture were identified through these contacts and data on the manufacturing process obtained. Where data were not available for a specific process the database was supplemented by data from the SimaPro database. Data have been collected based on the material components required for the various options, as shown in **Table D2.13.**

Table D2.13: The material components required for the six options

OPTION	COMPONENTS
1. Screen	6 mm drum screen at the storm outfall with 310 l/sec screening capacity
2. TBYF	Leaflets and questionnaires
3. Storage tank	Concrete tank (60m * 4m * 5m) to store 1100 m^3 of water
4. Source control	Plastic barrels, the number of barrels used with each sub-option (a, b, and c) are 626, 219 and 175 respectively.
5. Sewer rehabilitation	Replacing 300 mm, 450 mm and 600 mm pipes with different lengths for each sub-option. The pipes lengths for sub options (a, b, c, d, e and f) are: 144m, 108m, 87m, 54m, 42m and 30m respectively
6. Retrofit constricting WCs	Replacement of the toilet outlet connection with a smaller plastic pipe of 75 mm diameter and the cistern-flushing valve to reduce the flush volume to 3 l/flush.

Assembly/materials results

Figures D2.12a – D2.12d show SimaPro results for:

a) characterisation for assembly and material processing,

b) the normalised effects for an average European inhabitant over one year, the evaluation values, and

c) the indicator values.

All graphs show that the least environmental impact is that from implementing option 2 – running the TBYF public awareness campaign.

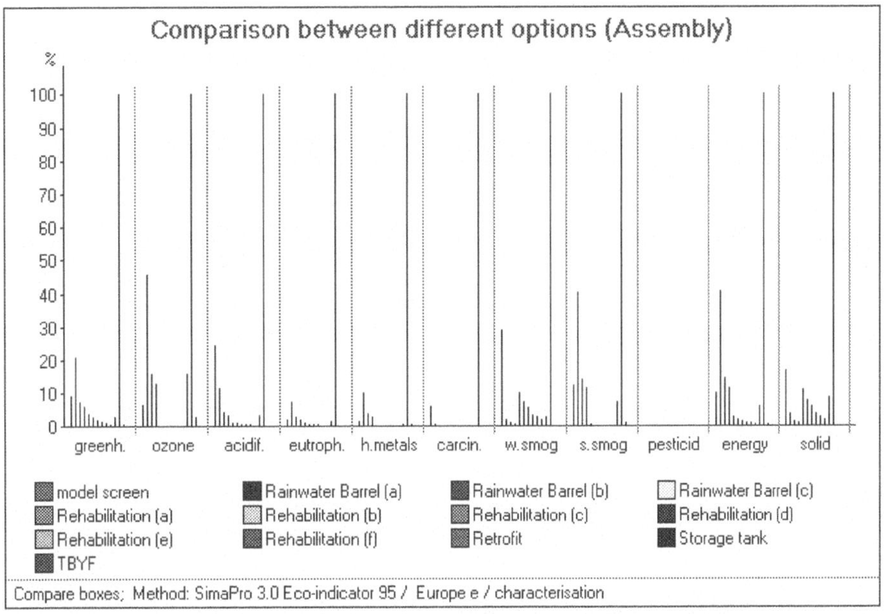

Figure D2.12a: Characterisation results for the options

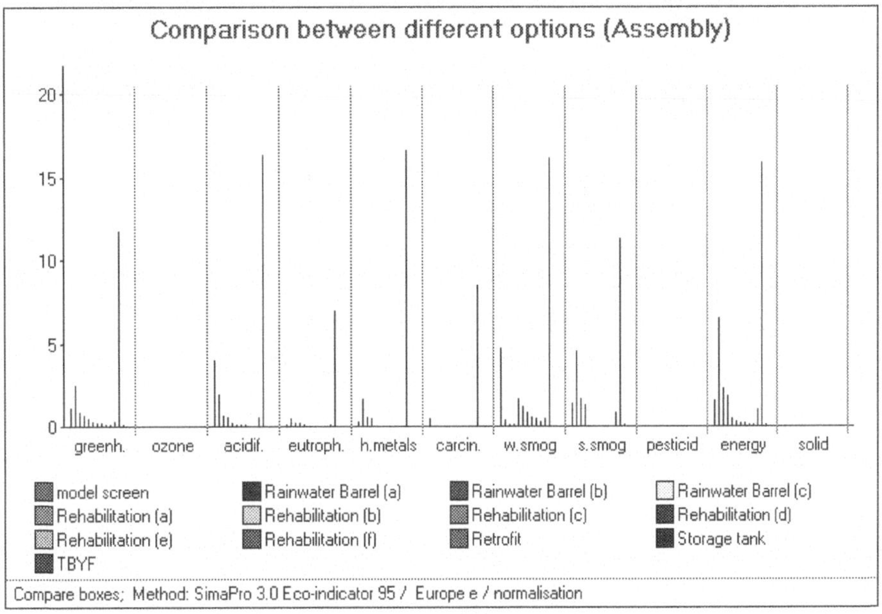

Figure D2.12b: Normalised environmental effect for the options

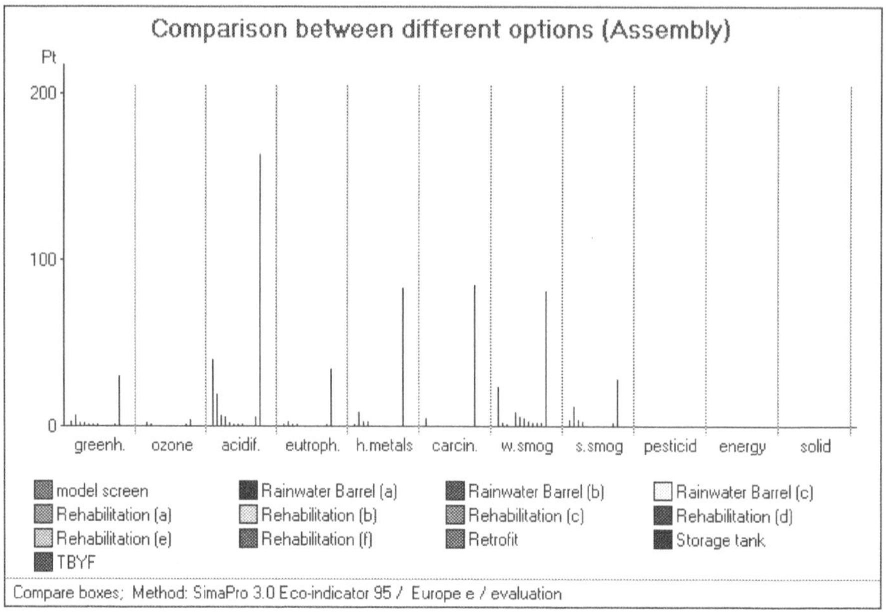

Figure D2.12c: Evaluation results for the studied options

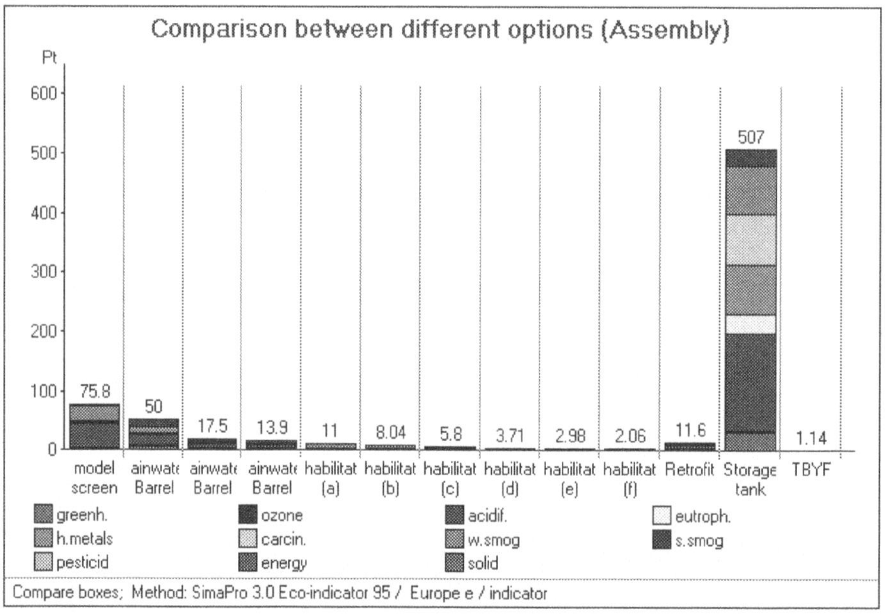

Figure D2.12d: Indicator results for the options

Emissions per kg gross solids reduction

In order to compare the projected total emissions from the options a functional unit has been selected. This is the *gross solids reduction during the option life cycle* (taken as two decades for the purpose of this case study).

Table D2.14 presents the emission values per kg of gross solids reduction for each option. The calculation of the gross solids reduction using the hydraulic model can be found in section D2.6.2.

Table D2.14: Emissions values per kg gross solids reduction

	Option		Gross Solids reduction in kg / 20 years	Green house kg CO_2 equ.	Acidification kg SO_4 equ.	NO_x kg	SO_2 kg	Energy MJ
1	6 mm Screen		2046.017	8.016	0.207	0.009	0.176	127.076
2	TBYF campaign		1416.453	0.569	0.003	0.000	0.000	11.790
3	Storage tank		1099.604	136.413	1.719	1.710	0.316	1282.279
4-a	Rainwater barrels	100%	959.016	18.248	0.111	0.009	0.016	535.966
4-b		35%	369.509	19.567	0.120	0.010	0.017	584.560
4-c		28%	303.404	16.018	0.097	0.008	0.014	471.320
5-a	Rehabilitation (% infiltration reduction)	100%	1262.132	4.683	0.022	0.012	0.001	59.265
5-b		72%	1056.150	4.090	0.019	0.010	0.001	51.887
5-c		42%	747.654	4.186	0.020	0.011	0.001	53.099
5-d		30%	595.259	3.444	0.017	0.009	0.001	43.678
5-e		18%	426.709	3.843	0.019	0.010	0.001	48.745
5-f		11%	269.049	4.200	0.021	0.010	0.001	53.522
6	Constricting WCs		2023.009	2.269	0.030	0.015	0.003	73.158

The results presented in **Table D2.14** show that the highest emissions and energy use per kg gross solids reduction is from option 3 (storage tank). Option 4b (source control – 35 % rainwater barrel installation) shows the second highest values for CO_2 and energy use per kg of gross solids reduction. **Figures D2.13a – D2.13d** present the total emissions of CO_2, NOx, SO_2 and SO_4 per kg of gross solids reduction for all of the options. **Figure D2.14** presents the total energy use for each of the options.

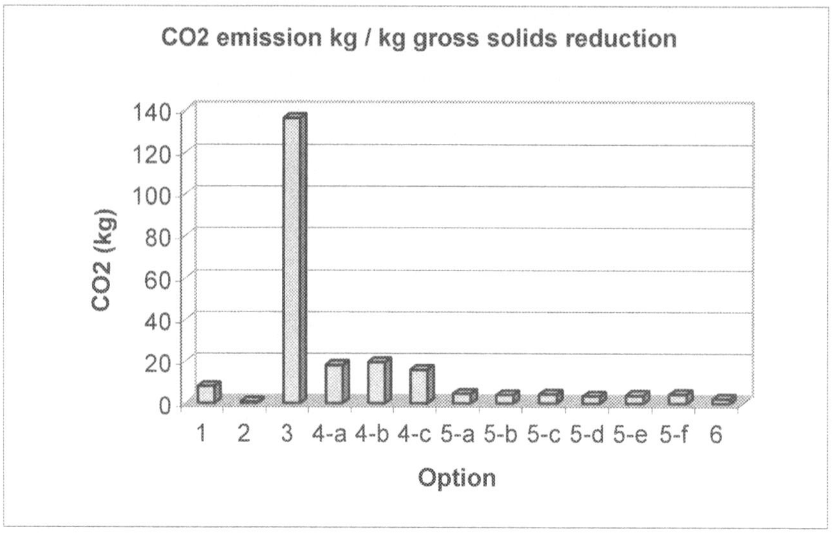

Figure D2.13a: CO2 emissions per kg gross solids reduction

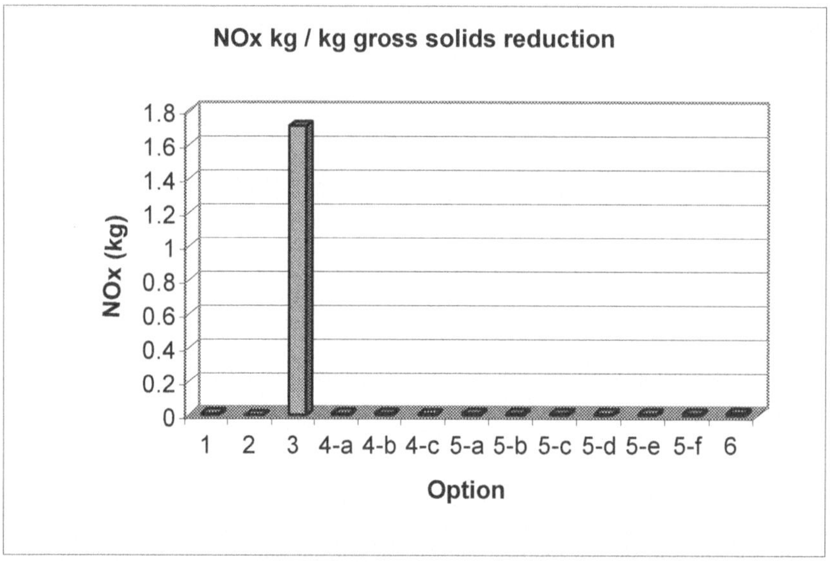

Figure D2.13b: NOx emissions per kg gross solids reduction

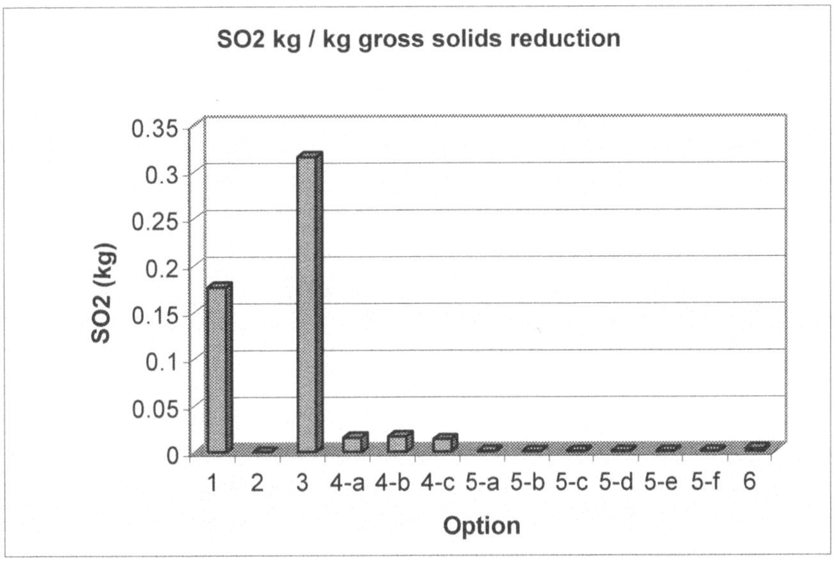

Figure D2.13c: SO2 emissions per kg gross solids reduction

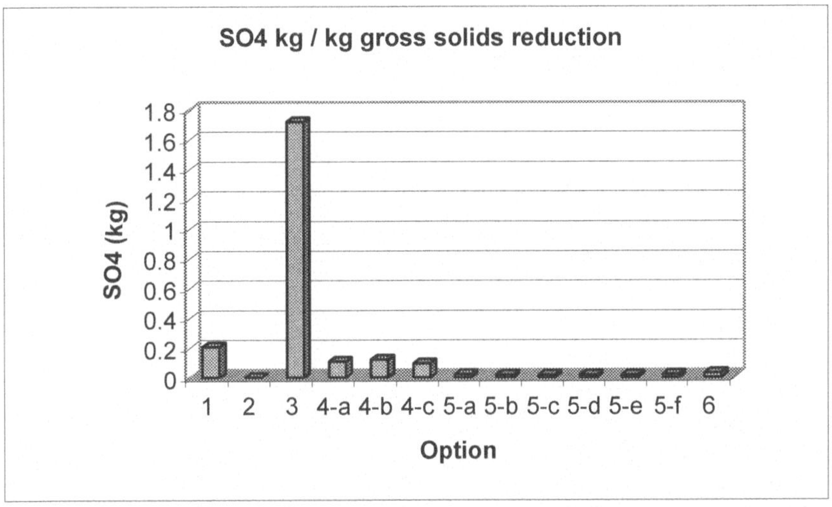

Figure D2.13d: SO4 emissions per kg gross solids reduction

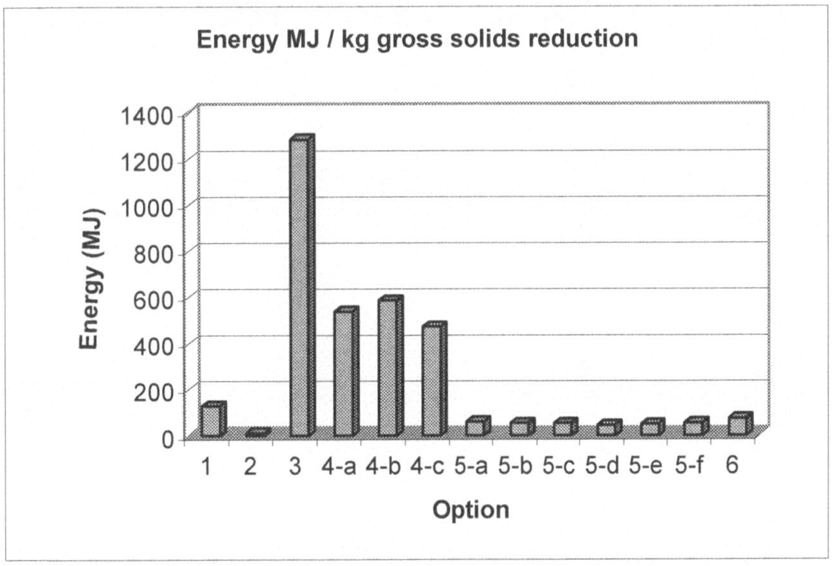

Figure D2.14: Energy use per kg gross solids reduction

EV3: CSO/SO discharges to the environment

This final environmental criterion was determined through a modelling approach that simulated the dry weather flow and rainfall-runoff-overflow transformation in the catchment. Model derivation involved the following:

- StormPac model: A one-year rainfall time-series dataset was been generated for the catchment using the StormPac package. The total volume of rainfall for the year was 842246 cu.m
- InfoWorks model: An InfoWorks model has been built for the catchment to simulate the dry and wet weather flow in the combined sewer system using the rainfall time-series data generated by StormPac. Velocity and depth in each pipe were determined. This was used to simulate storms for the 6 core options.

Flow volumes

Using the modelling approach outlined above, it was found that the internal CSO did not spill with the one-year complete time series simulation. **Table D2.15** shows the yearly flow simulation results for the storm outfall, treatment works and the combined total flow, total rainfall, total runoff and roof runoff for all the options studied.

The options tested were:

1. Install 6 mm screens on WwTW storm overflow and internal CSO.
2. 'Think Before You Flush' campaigns (with option 1).
3. Install an in-sewer storage tank
4. Retrofit stormwater source control (rainwater barrels) with the following degrees of disconnection:
    a. 100% roof area controlled
    b. 35% roof area controlled
    c. 28% roof area controlled
5. Sewer rehabilitation to exclude infiltration:
    a. Zero infiltration
    b. 72% infiltration reduction
    c. 42% infiltration reduction
    d. 30% infiltration reduction
    e. 18% infiltration reduction
    f. 11% infiltration reduction
6. Retrofit constricting WCs
    a. The amount of water flush/day/cap is assumed to be 45 l/day, 5 flushes/day/cap. The retrofit option will reduce the amount of flushed water to 15 l/day/cap, and reduce the gross solids flushed by 94%.

Table D2.15: Flow data for the different options for the catchment for a one-year simulation

Option	Storm outfall m^3/year	Flow to the WwTW m^3/year	Total flow m^3/year	DWF (inc. infiltration) m^3/yr	Total storm runoff (inc. roof) m^3/year	Roof runoff m^3/year
Baseline	58564.23	896208.3	954772.6	860496.6	94276	29344
1-2	58564.23	896208.3	954772.6	860496.6	94276	29344
3	28062.68	925777.5	953840.2	859564.2	94276	29344
4_a	32395.91	893877	926272.9	861340.9	64932	0
4_b	48481.28	896754	945235.3	861746.3	83489	18557
4_c	50408.63	896878.7	947287.3	861808.3	85479	20547
5_a	24765.82	180248.8	205014.7	110738.7	94276	29344
5_b	30360.58	389112.2	419472.8	325196.8	94276	29344
5_c	38607.65	601768.3	640375.9	546099.9	94276	29344
5_d	42699.34	692277.3	734976.6	640700.6	94276	29344
5_e	47238.78	776077.4	823316.2	729040.2	94276	29344
5_f	51398.19	822486.1	873884.3	779608.3	94276	29344
6	56626.74	881579.8	938206.6	843930.6	94276	29344

D2.6.3 Social criteria

S1 & S2: Acceptability to Stakeholders & Perceived Impact on the Environment

These two criteria were determined using a door-to-door survey conducted in the town. All surveys involved two members of the SWARD team, who were equipped with identity cards. Most people approached were happy to take part in the interview, and could remember the TBYF campaign that had previously been run in their area. After an introduction to the problem, and an interactive discussion about the respondents' perceptions of SW escape and their attitude to flushing, the logistics of the six options were briefly explained. Glossy coloured cards were produced to illustrate the six different methods of dealing with SW (see **Figures D2.15a – D2.15f**). The three alternative approaches to dealing with the problem - end of pipe, habit change and flow reduction – were also outlined and any questions answered.

All respondents were asked to rate the six options of tackling SW using a range of 1-5.

- For perceived impact on the environment 1 meant *very negative impact* and 5 meant *very positive* impact.
- For acceptability 1 meant not at all acceptable and 5 meant completely acceptable to the respondent.

The mean and standard deviation for each criterion were determined and are presented **in Tables D2.16 and D2.17**. The maximum and minimum raw scores for the criteria were calculated and the mean scores turned in to percentages accordingly, these figures are also reported in the tables.

Table D2.16: Mean, standard deviation and percentage scores for Criterion S1

S1: PERCEIVED IMPACT ON THE ENVIRONMENT	MEAN	SD	%
1. Install 6 mm screens	3.44	1.15	68.8
2. TBYF campaign	4.13	0.89	82.5
3. Install sewer storage	3.38	1.63	67.5
4. Source control (rainwater barrels)	3.31	1.08	66.3
5. Sewer maintenance	3.88	1.20	77.5
6. Retrofit constricting WCs	3.75	1.53	75.0

Table D2.17: Mean, standard deviation and percentage scores for Criterion S2

S2: ACCEPTABILITY TO STAKEHOLDERS	MEAN	SD	%
1. Install 6 mm screens	3.75	1.18	75.0
2. TBYF campaign	4.44	0.81	88.8
3. Install sewer storage	3.50	1.46	70.0
4. Source control (rainwater barrels)	3.06	1.39	61.3
5. Sewer maintenance	4.56	0.51	91.3
6. Retrofit constricting WCs	3.63	1.86	72.5

Table D2.16 shows that the least positive environmental impact was thought to be that which would arise from the installation of source control (rainwater barrels), whereas the TBYF public awareness campaign was thought potentially to have the most positive environmental impact. There was also the highest level of agreement amongst respondents about the public awareness campaign option. **Table D2.17** illustrates how the most acceptable option to those interviewed was the maintenance of the town's sewers in order to reduce infiltration to the system. There was also the highest level of agreement amongst the respondents about this option. The widest range of opinions was over option 6, the retrofitting of constricting WCs for both criteria examined. Respondents appeared to be either very positive or very negative about this option, with little neutral feeling expressed. Some additional comments on the options are outlined in **Table D2.18**.

Table 18: Observations and additional comments: door-to-door survey

Option	Additional observations and comments
1. Install 6 mm screens	▪ Most people were not aware that screens were currently available and used routinely. ▪ Many respondents enquired as to the final destination of the waste after it is removed into the skip.
2. TBYF campaign	▪ All respondents (except one, who was not living in the area at the time) could remember the TBYF campaign being run in their town. ▪ Two mentioned (without prompting) that the campaign had made a big impact on their behaviour. The majority said that the campaign had had some effect on them when asked, but voiced the opinion that "most people" (i.e. not them) do not care about these types of issue. ▪ "Educating the public is the best option, but it is very difficult. People do not listen – just look at the campaigns for not smoking, eating fat etc."
3. Install sewer storage	▪ Many respondents seemed to be very keen on this option. ▪ Several people mentioned the cost and disruption, but they thought that in the long term this was the easiest and most positive option for the town.
4. Source control	▪ People were generally very positive or very negative about source control. ▪ Two respondents had concerns about child safety if rainwater barrels were installed, whether these structures were covered or not. ▪ There were also some concerns as to where the water might go, people asked whether their lawn would get soggy?
5. Sewer maintenance	▪ Most people thought that sewer maintenance should be done routinely. ▪ There were many concerns voiced about the problem being widespread, although there was some confusion between water supply pipes and sewers. ▪ One respondent commented that they should maintain the (water supply) pipes to prevent pipe bursts, and thought that these bursts may also contribute to the excess water in the sewer system.
6. Retrofit constricting WCs	▪ People were generally very positive or very negative about this option. ▪ One respondent had tried low flush toilets and subsequently had them removed as they claimed that they ended up flushing several times to remove any (human) waste. They also commented that their mother was experiencing the same difficulty and was going to have her low flush toilet removed. ▪ Another respondent thought that the toilets would be a nuisance. ▪ Several respondents mentioned the cost - "They should be spending money on maintaining the pipes and treatment works, not installing this sort of thing". ▪ Two respondents thought that the toilets were a very good idea as long as someone else paid them for. ▪ One thought that the toilet was a good idea to conserve water generally.

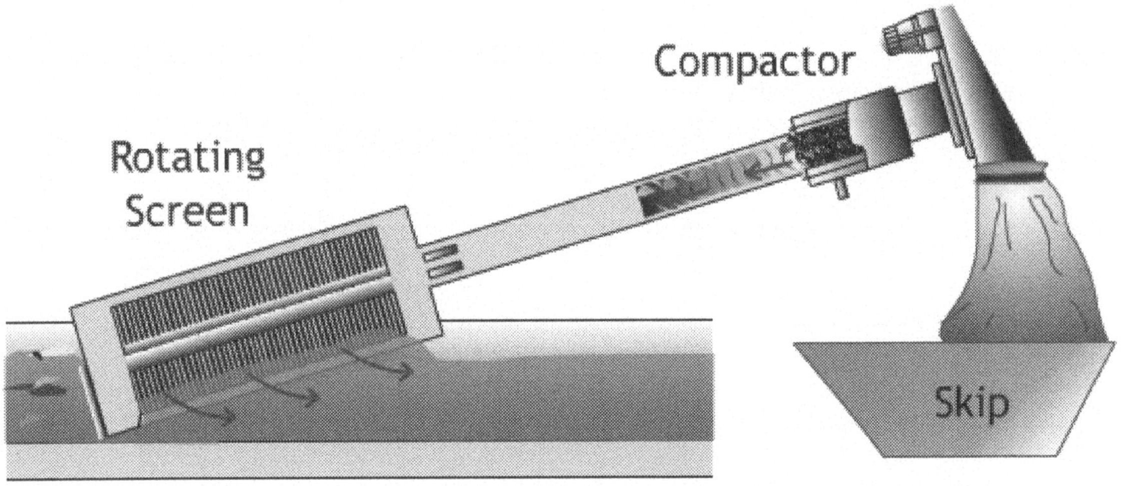

Figure D2.15a: Survey flash card 1. **Electrical Screen:** Sanitary waste is screened and removed at the treatment works using an electrically powered rotating screen

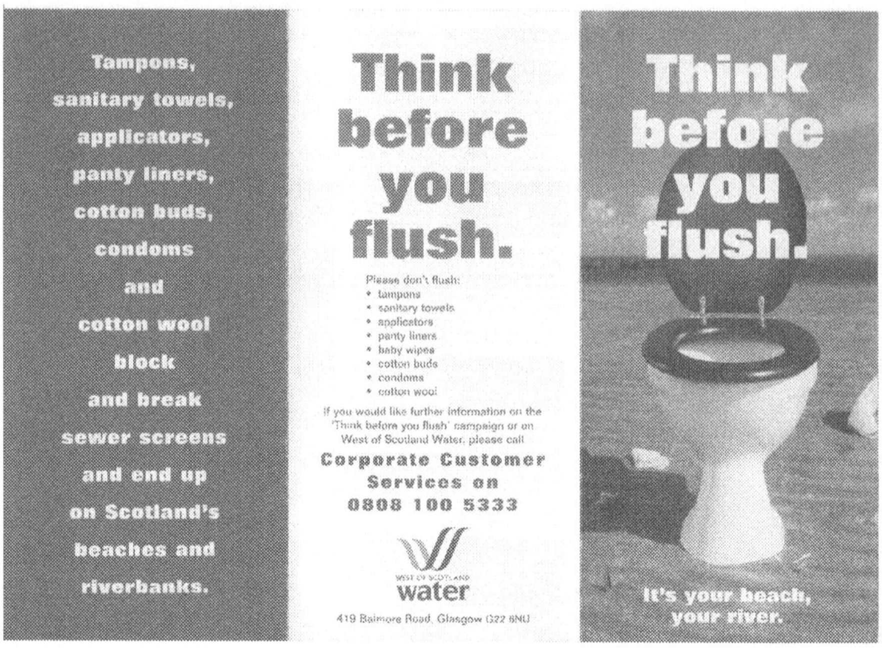

Figure D2.15b: Survey flash card 2. **Think Before You Flush:** Running a campaign in a community which encourages people to stop flushing sanitary waste items

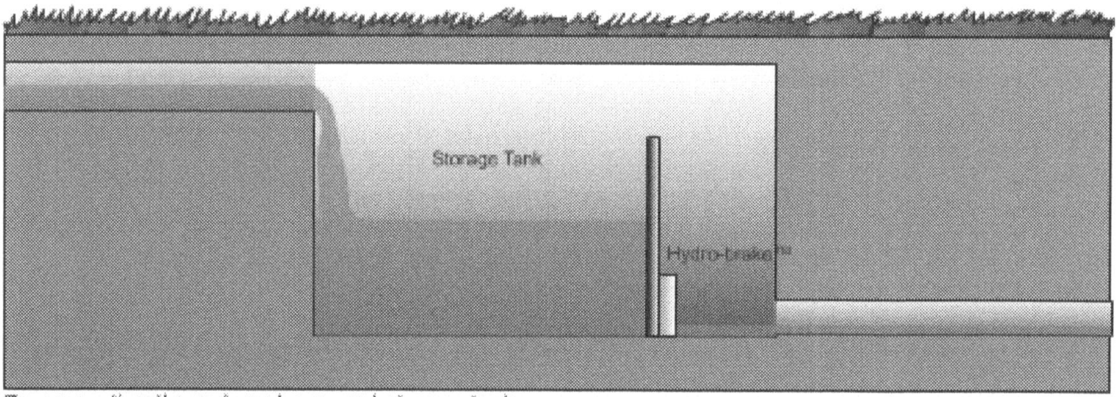

Cross-section through underground storage tank

Figure D2.15c: Survey flash card 3. **Storage Tank:** Collects water in the sewer during heavy rainfall. Flows are lower and this stops sanitary waste from being discharged to the environment via storm overflows.

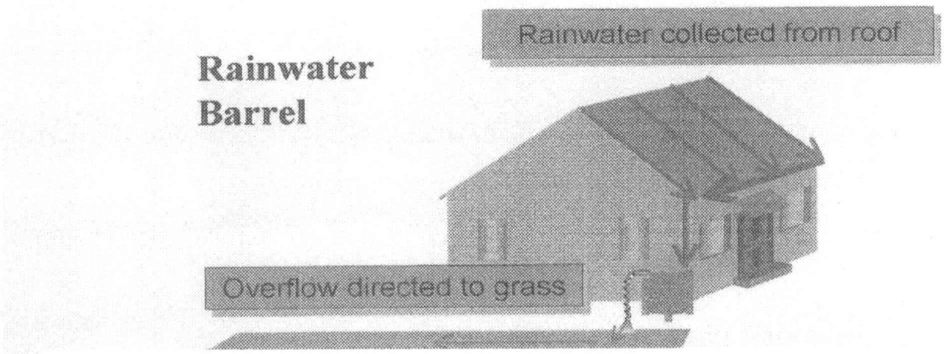

Rainwater barrels. These stop water flowing directly into the sewer system. Flows are lower and this stops sanitary waste from being discharged to the environment via storm overflows.

Figure D2.15d: Survey flash card 4

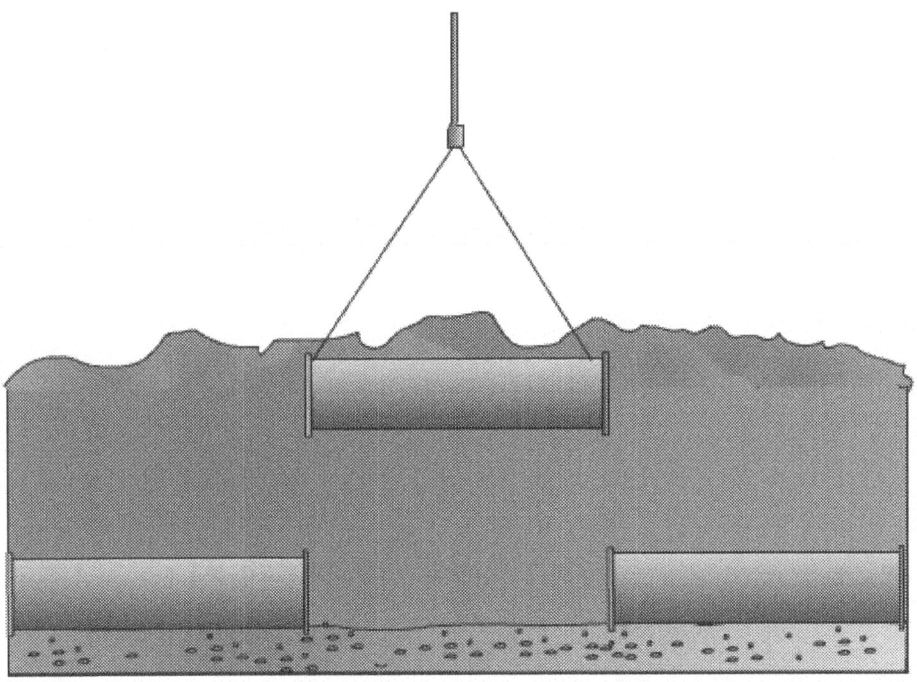

Figure D2.15e: Survey flash card 5. ***Sewer maintenance:*** *Sewer maintenance prevents stream water entering the sewer through cracks and displaced joints. Flows are lower and this stops sanitary waste from being discharged to the environment via storm overflows*

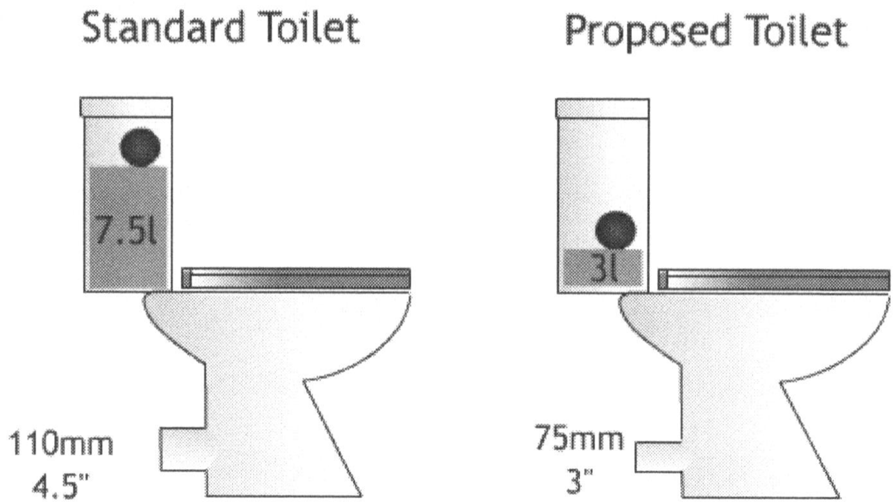

Figure D2.15f: Survey flash card 6. ***Constricting WCs**: Low flush toilets with a smaller outlet pipe restrict the flushing of sanitary waste and reduce water consumption*

S3: Participation and responsibility

The level of participation and responsibility required for each of the six options was determined through consultation and debate involving water authority personnel, social and behavioural experts and other academic members of the SWARD team. Options such as screen installation, in-sewer storage and sewer rehabilitation obviously involve no behavioural change on the part of the local population. Source control and retrofit options involve increased public participation in action for more sustainable behaviour, whereas the TBYF option encourages a fundamental shift in attitudes and behaviour in regard to SW disposal.

D2.6.4 Technical criteria

T1: Sanitary Waste Escape

Gross solids entering the sewerage network can include items such as: plastics and rubber; tampons, panty liners and sanitary towels (including backing strips); cotton buds; facial or baby wipes; cigarette ends; and cotton wool. In order to calculate the SW mass balance, i.e. items flushed via WC, retained in the sewerage system, and escaping to the environment through the storm outfalls, the following procedure was carried out:

Sanitary waste entering the system

Current scenario A statistical approach using a Poisson distribution was used to estimate the distribution of the gross solids (tampons and sanitary towels) entering the sewerage system at the present time. The data was based on population, number of women and their age, and number of babies in the town. Data regarding the average number of items used per person per day was available from the surveys conducted prior to the TBYF campaign. The population data presented in **Table D2.19** were obtained from the 1991 census.

Table 19: Population data: 1991 census

Age (years)	Total persons
Children 0-4	69
Female 12-17	35
Female 18-59	426

The distribution of the gross solids entering the system for the catchment is shown in Table D2.20. The distribution principle is based on the social survey data results before the Think Before You Flush (TBYF) campaign [(4)].

Table D2.20: Gross solids average use distribution profile for the catchment

	Applicators	Tampon	Baby wipes	Cotton buds	Panty liners	Sanitary towels
No/day/user	4-5	4-5	5	1-2	4-5	4-5
No/month/user	20-30	20-30	150	20-40	20-30	20-30
No. of days/month	5	5	30	20	5	5
No. of women/day	23	23	-	-	7	10
Total No. of woman/month	195	195	-	-	63	85
No. of baby	-	-	38	-	-	-
No. of Cotton buds users				303		

The assumptions in the above distribution are based on 20% of the total population flushing cotton buds. The number of women flushing SW is assumed to be 75% of the total female population. The estimated number of women flushing applicators is 50% of the tampon flushers from the survey data. **Table D2.21** show the estimated number and weight of gross solids flush via WC for the catchment. The estimated weight has been used with all options except for TBYF (2) and constricting WC (6) options.

The amount of screened solids collected in the treatment works was recorded as 11.2 tonnes/year in 1998-1999. The amount of screened solids calculated from the model (option 1) using the gross solids assumption is 10.8 tonnes/year (10865.535 kg). The difference between the modelled value and recorded value is 3.5 %. This can be considered as the weight of toilet paper and other miscellaneous items captured by the screens.

TableD2. 21: Estimated number and weight of gross solids flush via WC for the catchment

	Applicators	Tampons	Baby wipes	Cotton buds	Panty liners	Sanitary towels
No/day/user	5	5	5	1-2	5	5
No/month/user	25	25	150	20	25	25
No. of days/month	5	5	30	20	5	5
No. of women/day	23	23	-	-	7	10
Total No. of woman/month	98	195	-	-	63	85
No. of babies	-	-	38	-	-	-
No. of cotton bud users				303		
Total No/month	2450	4875	5700	6060	1575	2125
Total No/year	29400	58500	68400	72720	18900	25500
Ave wt (g)/item*	3.328	33.296	19.571	0.198	60.539	247.947
Ave wt (g)/year	97836.97	1947843	1338656	14362.2	1144188	6322649
Ave wt (kg)/year	97.837	1947.843	1338.656	14.362	1144.188	6322.649
Total Ave wt (kg)/year*				10865.535		

* wet weight

Think before you flush (TBYF) The values stated below in **Table D2.22** are for the sanitary waste distribution for disposal of different items obtained from the TBYF campaign social survey data. The survey data from running the campaign (option 2) shows a reduction of 65% in the total amount of gross solids entering the system.

Table D2.22: Flushing profile for 'Think Before You Flush' Campaign: Option 2

	Applicators	Tampons	Baby wipes	Cotton buds	Panty liners	Sanitary towels
No/year	7800	15240	63360	22680	5040	6600
Ave wt (g)/item*	3.328	33.296	19.571	0.198	60.539	247.947
Ave wt (g)/year	25958.40	507431.04	1240018.5 6	4490.64	305116.56	1636450.20
Ave wt (kg)/year	25.957	507.438	1240.019	4.479	305.117	1636.450
Total Ave wet wt kg/year			3719.47			

Retrofitting constricting WCs For the retrofit option (6) it is assumed that although sanitary towels will not be flushed, 15% of certain other sanitary waste items will still be flushed, resulting in 681.435 kg of gross solids entering the system per year (**Table D2.23**). The data shows that the lowest amount of GS escape to the environment through the storm outfall is for option 6.

Table D2.23: Flushing profile for retrofitting constricting WCs option

	Applicators	Tampons	Baby wipes	Cotton buds	Panty liners
No/year	4410	8775	10260	10908	2835
Ave wt (g)/item*	3.328	33.296	19.571	0.198	60.539
Ave wt (g)/year	14676.48	292172.40	200798.46	2159.78	171628.07
Ave wt (kg)/year	14.676	292.172	200.798	2.160	171.628
Total Ave wt (kg)/year			681.435		

Gross solids calculation methodology

The method used to calculate the total amount of gross solids escape to the environment through the storm outfall is based on the flow ratio of the storm overflow and the treatment works flow to the total flow. The total gross solids entering the system over a one-year period are assumed to be equal to the total amount of gross solids leaving the system.

The storm pump flow rate is almost 9 times higher than the flow rate of the treatment works pump. The gross solids calculation using this method is based on the working duration of the storm pumps and its flow ratio to the total flow. The data utilised in the calculation are the discharge rate from the storm pumps and the total discharge from the storm outfall pumps. The pump working duration is then calculated. Where the treatment works pump capacity is known, flow to the treatment works while the storm pumps are working can be calculated. From the previous calculation, total flow during the spill duration can be estimated by summing both flows. The ratio of the storm outfall flow and the treatment works flow to the total flow can then be calculated. The gross solids will be proportionally distributed between the treatment works and storm outfall according to the flow ratio.

For the storm outfall pump duration calculation, only the working duration of the first stop-start pump has been considered. This is because the second pump will only start working in parallel with the first pump. This is to avoid increasing the working duration of the treatment works pumps, which will consequently increase the flow to the treatment works and its ratio to the total flow. The methodological steps to simulate the gross solids movement in sewer systems are outlined below:

1. Storm pump capacity, 150 l/sec each (1)
2. Treatment works pump capacity, 18 l/sec each (2)
3. Total discharge of storm pumps for each option (3). This value has been calculated using the InfoWorks model.
4. Storm pump working duration = (3) / (1): For this step only the first working pump discharge is considered, which is pump 4.
5. Flow to the treatment works during the total storm duration = (4) * (2)
6. Total flow = (3) + (5)
7. Storm pumps flow proportion to the total flow = (3) / (6)
8. Treatment works flow proportion = (5) / (6)
9. To calculate the gross solids, record the amount of gross solids/ year for each option.
10. Amount of gross solids/ day = (9)/(12*30)
11. Total gross solids existing in the network during the working time of the storm pumps = (10) * (4)
12. Gross solids leaving the systems with the storm pumps = (11) * (7)
13. Gross solids to the treatment works during the storm duration = (11) * (8)
14. Total gross solids to the treatment works during the year (storm + dry weather) = (9)-(12)

Table D2.24 shows the gross solids distribution to the treatment works and storm outfall according to the flow ratio during storm overflow. It was assumed that the screen efficiency for option 1 is 95%. The amount of gross solids reduction per annum was calculated for each option based on present average gross solids flush via the WC, 10865.535 kg/year wet weight.

Figure D2.16 shows the total gross solids distribution for the different options. The results show that the lowest value for gross solids ratio for the storm outfall is obtained by option 6 - retrofit/constricting the WC, followed by the TBYF campaign option (option 2). Both options involve reducing the amount of SW disposed of via the WC. The results also show the different distribution for gross solids escaping to the environment for the remaining options. The amount of gross solids leaving through the storm outfall for the storage tank (option 3) show slightly lower values than the rainwater barrel source control (option 4a).

Table D2.24: Gross solids distribution in the system outfalls

Option			GS to Storm outfall kg/year	GS to Treatment works kg/year	Total solids kg/year	GS reduction kg/year
Baseline scenario			107.685	10757.850	10865.540	-
1	Screen (95% efficiency)		107.685	10757.850	10865.540	102.301
2	TBYF		36.862	3682.598	3719.460	70.823
3	Storage tank		52.705	10812.830	10865.540	54.980
4-a	Rainwater Barrel	100%	59.734	10805.801	10865.540	47.951
4-b		35%	89.210	10776.325	10865.540	18.475
4-c		28%	92.515	10773.020	10865.540	15.170
5-a	Infiltration	0%	44.579	10820.956	10865.540	63.107
5-b		72%	54.878	10810.657	10865.540	52.808
5-c		42%	70.302	10795.233	10865.540	37.383
5-d		30%	77.922	10787.613	10865.540	29.763
5-e		18%	86.350	10779.185	10865.540	21.335
5-f		11%	94.233	10771.302	10865.540	13.452
6	Retrofit		6.535	674.898	681.433	101.150

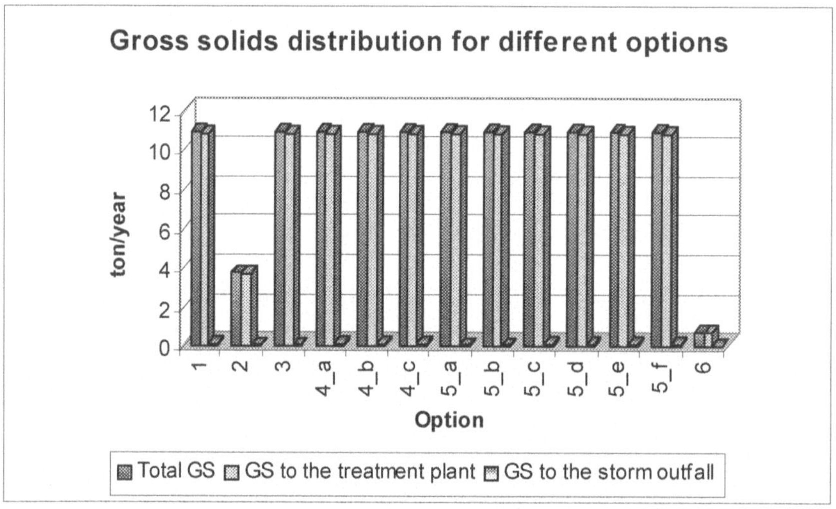

Figure D2.16: Total gross solids distribution

Figure D2.17 shows the gross solids reaching the storm outfall for the options. Option 1, the end-of-pipe solution, has the highest amount of gross solids. The remaining options result in either a reduction in the amount of SW disposed of via WC (options 2 and 6) or a reduction in storm overflows from the system (options 3, 4 and 5).

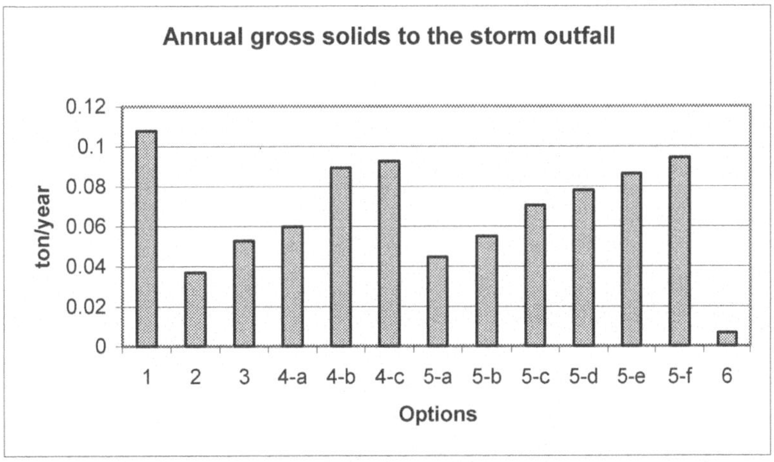

Figure D2.17: Annual gross solids to the storm outfall

Figure D2.18 illustrates the gross solids reduction for each option per annum. The 6 mm screen option (1) has the highest GS reduction value followed by the sewer rehabilitation (5-a) option.

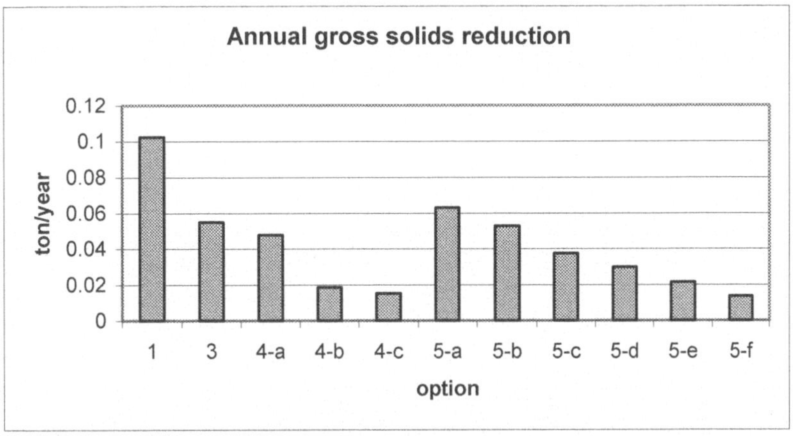

Figure D2.18: Gross solids reduction per annum

Figure D2.19 shows the amount of gross solids escape via the storm overflow for each of the options. The results show that the screen option (1) has the lowest value followed by options 6, 2 and 3. Options 4 and 5 have different values according to the percentage of control initiated.

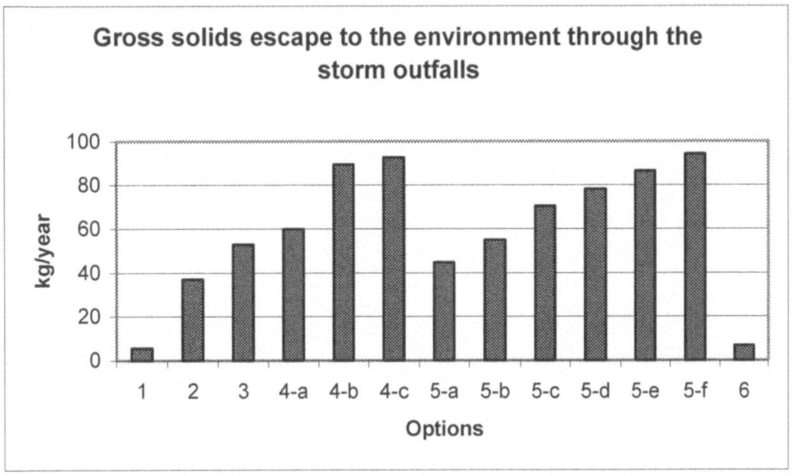

Figure D2.19: Gross solids escape through the storm outfall to the environment

T2: Sanitary Waste Transport In Sewer

It was originally intended that this criterion would be modelled using InfoWorks (mass balance approach) and Matlab/Simulink, however the detailed modelling approach to determine this is still under development as part of an ongoing PhD research project. The criterion is therefore expressed qualitatively. Discussions were held between academic members of the SWARD team to reach a consensus on the effects on in-sewer transport and deposition. These took account of the effects on gross solids entering the sewer and on the hydraulic regime of each of the six options. Examining the gross solids and hydraulic data produced from InfoWorks for the criteria EV3 'CSO Discharge' and T1 'Sanitary Waste Escape', also helped a best estimate to be made.

- Option 1 (installing 6 mm screen) would not result in any change in entry, in-sewer transport, deposition and re-suspension of SW.
- Options 2 (TBYF campaign) and 6 (retrofit constricting WCs) would result in a significant reduction in the amount of SW entering and moving through the system.
- Option 3 (install sewer storage) will not reduce the amount of SW entering the sewer system, and may result in the deposition of SW in the storage tank. Subsequent heavy precipitation conditions may then lead to the erosion and washout of these solids, leading to the problems at the WwTW such as the sudden blinding of screens.
- Option 4 (source control – rainwater barrels) again will not lead to the reduction of SW into the system. There may be some re-suspension of temporarily deposited SW in times of increased discharge. However, this may not be a serious problem as wet weather flows will be lower.
- Option 5 (sewer rehabilitation) will have no effect on the amount of SW entering the system, but as a result of reduced infiltration and thus in-sewer flow, may lead to significant deposition and subsequent re-suspension of SW during wet weather flows.

T3 & T4: Risk of failure to provide service & Flexibility and adaptability

The data for these criteria were determined through discussion with water authority personnel and given qualitative values. Values for the criterion 'Risk of failure to provide service' ranged from 'very low' for option 1 (6 mm screen) to 'very high' for option 6 (retrofit constricting WCs), whereas the values given by the decision-maker for the flexibility and adaptability of the options ranged from high for option 4 (source control – rainwater barrels) to low for options 3, 5 and 6 (in-sewer storage, sewer rehabilitation and retrofit constricting WCs).

Table D2.25 presents all of the data collected and information generated for the 16 case study criteria.

Criteria	OPTIONS						
	1. SCREEN	2. TBYF	3. STORAGE	4. SOURCE CONTROL	5. REHABILITATION	6. RETROFIT	Units
Sanitary Waste Escape	5.38	36.86	52.7	89.21	70.3	6.54	kg/yr
SW Transport	low	low	medium	Medium	high	medium	Qualitative
Risk of failure	very low	high	low	Medium	low	very high	Qualitative
Flexibility and adaptability	medium	medium	low	High	low	low	Qualitative
Capital cost	68000	10000	70000	3925	39000	32446	£
Operational cost	628.67	800	0	0	0	800	£/yr
Maintenance cost	3167	0	680	0	0	0	£/yr
Financial risk exposure	medium	low	high	Low	low	high	Qualitative
Acceptability	68.8	85.9	62.5	51.6	89.1	65.6	%
Perceived impact on the environment	60.9	78.1	59.4	57.8	71.9	68.8	%
Participation and responsibility	low	high	low	High	low	high	Qualitative
Energy use	2.6E+05	1.67E+04	1.41E+06	2.16E+05	3.97E+04	1.48E+05	MJ
Impact on air - CO_2	1.64E+04	806	1.5E+05	7.23E+03	3.13E+03	4.59E+03	kg
Impact on air – Nox	18.9	0	1.88E+03	3.63	7.94	31.1	kg
Impact on air – SO_4	423	3.75	1.89E+03	44.4	15	60	kg
CSO Discharge	58564.23	58564.23	28062.68	48480.28	38607.65	56626.74	m^3/yr

Table 25: Collated results for all criteria

D2.7 SWARD PHASE 5: DATA ANALYSIS

D2.7.1 MCDA Analysis Methods

Methods Adopted and Data Preparation

Three MCDA methods were used in the case study: ELECTRE, PROMETHEE, and SMART. These methods were selected for the following reasons:

- PROMETHEE and ELECTRE III are non trade-off methods that should be used in strategic, complex decision-making problems, that have far reaching consequences in time or involve substantial capital costs. One of the differences between these two methods is that PROMETHEE has six predefined value functions. As such its use is recommended only if the criteria selected can be modelled by one of these predefined value functions. This confers the methodology a more restrictive use.
- ELECTRE III, on the other hand, has no predefined value functions. When using this methodology one has to model these functions for each criterion individually. The methodology is more complex and accurate and can deal with a very high number of variables (criteria and options). By using a veto threshold, unacceptable values for criteria or unacceptable options are eliminated from the analysis.
- SMART is a very simple methodology recommended in every day decision-making problems of a multi-criteria nature. The major weakness of this model is that it is a trade-off method and can be used for the analysis of up to 16 criteria at the most. Its compensatory nature can hide unacceptable values for criteria.

These three methods were used to help the decision-maker find a preferred option among the six identified, taking into account the 16 criteria. The importance coefficients (weights) of the four groups of criteria, of the 16 criteria individually, and of the criteria in each group, were determined in the workshop described in Section 5 of this case study.

A meeting was held with a decision-maker to obtain data on his value system. Members of the SWARD team identified the preference, indifference and veto threshold values of the decision-maker with regard to the data obtained for the 16 criteria for each of the options. These values were needed for input into the three MCA models.

For PROMETHEE:

The decision-maker was asked:

- Which of the six possible value function types defined and specified for use in the PROMETHEE methodology/software best corresponded to each of the 16 criteria. The six value function types can be seen in Appendix D of the SWARD Guidebook.
- To determine the two discrimination thresholds, namely: the preference threshold (that indicates the value at which there is a significant preference for an option) and the indifference threshold preference (that indicates the value at which there is no preference between the compared options).

For ELECTRE:

Before the meeting with the decision-maker, members of the SWARD team had modelled the value functions for each of the 16 criteria. These were of a linear or curved shape in the case of the quantitative criteria and a stepped shape for the qualitative criteria. The decision-maker was asked:

- Whether the value functions accurately modelled the 16 criteria considered. Alterations were made where required.
- The preference thresholds
- The indifference thresholds

- The veto threshold (indicating the point at which any value above or below is totally unacceptable to the decision-maker).

For SMART:
No information was required from the decision maker for input into the SMART analysis. The data obtained from the workshop that was held with water industry personnel (see Section 5) was utilised to establish the SMART preference scale. For this purpose a score of 100 was associated with the most preferred option, with 0 associated with the least preferred. Value functions were established by using information gathered in the collaborators workshops and information given by the decision maker. Sketches were made of the likely shapes of the value functions by the case study team and these were presented to the decision-maker for the SW case study. The decision-maker then validated these value functions stating whether these were representative of his views and value system.

An example of a quantitative and a qualitative value function are shown below. The quantitative function shown is a curve that represents the relative value put on the reductions in SW discharge for each option as shown in **Table D2.25**. Values were generated for each option from this and appropriate value functions were also developed for each criterion.

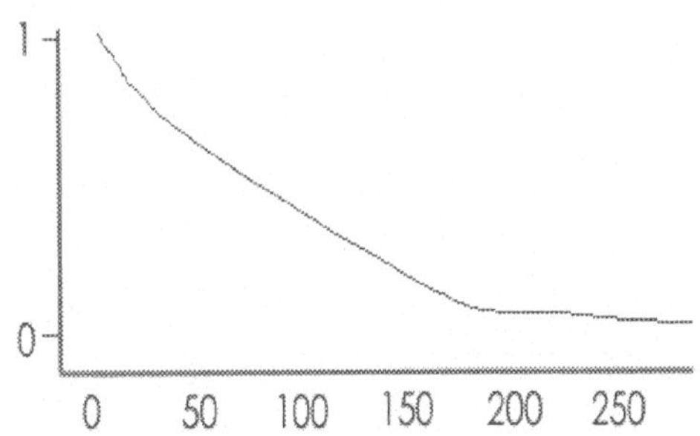

Figure D2.20: Value Function for Sanitary Waste escape

Figure D2.21 shows an example of a value function for a point 5 qualitative scale. For this criterion a value has been put on each scale from very low to very high. It can be seen that the steps between each scale can be different.

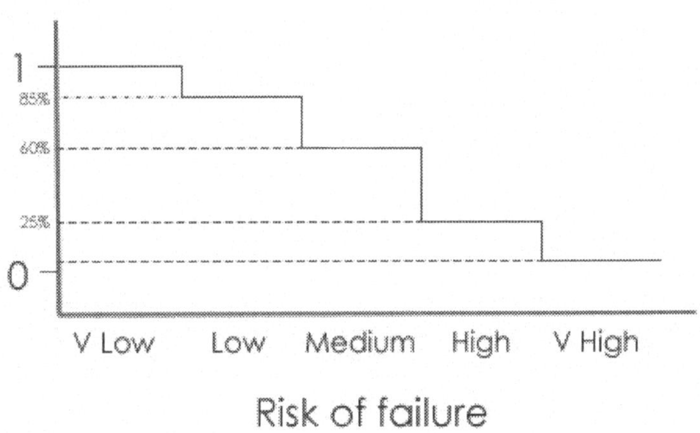

Figure D2.21: Value Function for sanitary waste escape.

In this section, the analyses used in PROMETHEE, ELECTRE, and SMART in the context of reducing the number of sanitary waste items escaping to the environment are presented. In each of the following analyses, the criterion and option codes were used as shown in **Tables D2.26 and D2.27**.

Table D2.26: List of option codes

Option	Code
Install 6mm Screens on overflows at WwTP (6mm already installed on CSO)	In6S
'Think Before You Flush' (TBYF) campaigns	TBYF
Install Flow Storage	InSt
Retrofit Stormwater Source Control	RSSC
Sewer Rehabilitation	SeRe
Retrofit Outlet Chokes on existing WCs and introduce these to new developments	ROC

Table D2.27: List of criteria codes

Criteria	**Code**
Economic	
Capital Cost	CC
Operational Cost	OC
Maintenance Cost	MC
Financial Risk Exposure	FRE
Social	
Acceptability (to Stakeholders)	AcSt
Perceived Impact on the Environment	PIE
Participation and Responsibility	PaRe
Environmental	
Sanitary Waste Escape	SWE
(Primary) Energy Use	EnUs
Impact on Air - CO_2	IoAC
Impact on Air - NO_x	IoAN
Impact on Air - SO_4	IoAS
CSO Discharge	CSOD
Technical	
Sanitary Waste Transport	SWT
Risk of Failure	RoF
Flexibility and Adaptability	FlAd

Weighting distributions of the criteria

The weighting distribution of the criteria was based on the collective prioritisation of the SWARD primary/generic criteria. The weighting of each criterion was determined by multiplying the prioritisation factor of each of the four main classification groups (as detailed in **Figure D2.10**) to which each criterion belongs to the inter-category importance of the same criterion. For example, it is shown elsewhere that the collective prioritisation of the four groups, economic, social, environmental, and technical, were determined to have 35%, 15%, 30%, and 20% importance respectively. Also, within the technical criteria, the inter-category importance of the 'Risk of Failure' (RoF) criteria was determined to be 45%. Thus the overall weighting of this criteria was determined to be $0.2 \times 0.45 = 0.09$. This was therefore given a weighting factor of 9. In this context, the weightings of all the criteria are detailed in **Table D2.28**, and were fed into the three decision-aiding analyses labelled as 'Normal' weights.

Table D2.28:Criteria 'normal' weightings

Criteria	'Normal' *Weighting*
CC	10.5
OC	10.5
MC	10.5
FRE	3.5
AcSt	4.5
PIE	7.5
PaRe	3
SWE	8
EnUs	3.5
IoAC	3.5
IoAN	3.5
IoAS	3.5
CSOD	8
SWT	4
RoF	9
FlAd	7
Total	**100**

PROMETHEE Analysis

The PROMETHEE procedure is based on pairwise comparisons. From the six predefined preference functions the decision-maker was asked to select the one he considered better modelled the criterion in question. For each criterion 0, 1 or 2 parameters have to be defined. The significance of these parameters is: *q* is a parameter (threshold) of indifference, *p* is a threshold of strict preference and *s* is in between. The input data and parameter settings can be seen **in Tables D2.29 and D2.30**. The PROMCALC software was used for the analysis. It is developed for real world applications and can perform up to 3600 evaluations

By introducing the data from **Table D2.30** into the software, a partial ranking was obtained using PROMETHEE I. The partial ranking is the intersection of positive and negative outranking flows. The positive (Φ+) and negative (Φ-) outranking flows are usually not identical. In this case the information of both outranking flows is consistent and may therefore be considered as reliable. The partial ranking is presented in **Figure D2.22**.

Table D2.29: PROMETHEE Input data

	SWE	SWT	RoF	FIAd	CC	OC	MC	FRE	AcSt	PIE	PaRe	EnUs	IoAC	IoAN	IoAS	CSOD
In6S	5.38	0.0000	0.0000	0.6000	68000.00	628.67	3167.00	0.3000	68.8000	60.9000	0.0000	260000	16400	18.90	423.00	58564.23
TBYF	36.86	0.0000	0.7500	0.6000	10000.00	800.00	0.00	0.0000	85.9000	78.1000	1.0000	16700	806	0.00	3.75	58564.23
InSt	52.70	0.5000	0.1500	0.0000	70000.00	0.00	680.00	1.0000	62.5000	59.4000	0.0000	1410000	150000	1880.00	1890.00	28062.68
RSSC	89.21	0.5000	0.4000	1.0000	3925.00	0.00	0.00	0.0000	51.6000	57.8000	1.0000	216000	7230	3.63	44.40	48480.28
SeRe	70.30	1.0000	0.1500	0.0000	39000.00	0.00	0.00	0.0000	89.1000	71.9000	0.0000	39700	3130	7.94	15.00	38607.65
ROC	6.54	0.5000	1.0000	0.0000	32446.00	0.00	800.00	1.0000	65.6000	68.8000	1.0000	148000	4590	31.10	60.00	56626.74

Table D2.30: PROMETHEE parameter settings

	SWE	SWT	RoF	FIAd	CC	OC	MC	FRE	AcSt	PIE	PaRe	EnUs	IoAC	IoAN	IoAS	CSOD
Function Type	III	II	IV	II	V	III	V	II	III	IV	II	V	II	II	II	V
Min/Max	Min	Min	Min	Max	Min	Min	Min	Min	Max	Max	Max	Min	Min	Min	Min	Min
P	19	2	0.3	2	20000	500	500	2	10	5	2	100000	2	2	2	15000
Q	1	0.3	0.2	0.39	10000	1	200	0.2	1	4	0.3	99000	50000	300	250	12000
S	1	1	1	1	1	1	1	1	1	1	1	1	1	1	1	1
Unit	kg/yr	Qualitative	Qualitative	Qualitative	£	£/yr	£/yr	Qualitative	%	%	Qualitative	MJ	kg	kg	kg	Litres/yr
Scale	Numerical	Low 0 Medium 0.5 High 1	V.Low 0 Low 0.15 Medium 0.4 High 0.75 V.High 1	Low 0 Medium 0.6 High 1	Numerical	Numerical	Numerical	Low 0 Medium 0.3 High 1	Numerical	Numerical	Low 0 High 1	Numerical	Numerical	Numerical	Numerical	Numerical
Weight	8	4	9	7	10.5	10.5	10.5	3.5	4.5	7.5	3	3.5	3.5	3.5	3.5	8

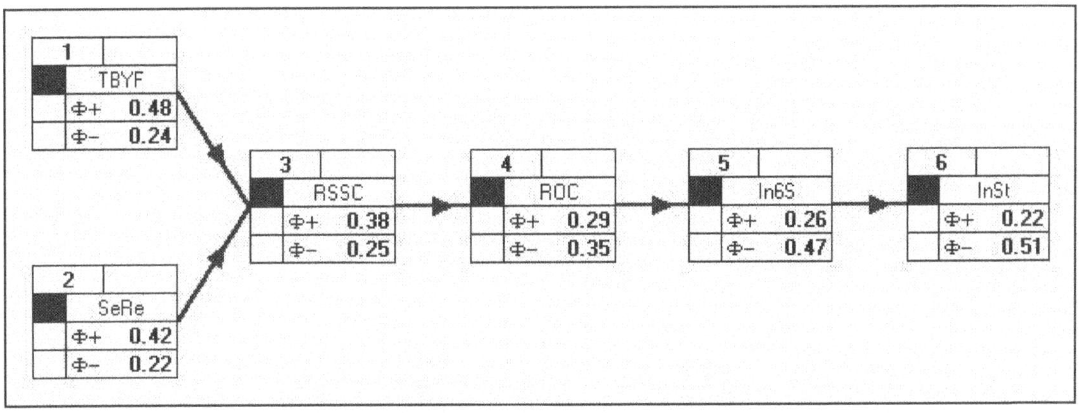

Figure D2.22: Partial Ranking (PROMETHEE I)

Figure D2.22 shows that TBYF scores 'well' on a set of criteria on which SeRe (sewer rehabilitation) is weak and vice-versa. As the information corresponding to both flows, is not consistent, it seems natural to consider the alternatives as incomparable. The partial ranking does not indicate which is better. It is up to the decision-maker to take his or her responsibility. This is why the decision-maker often requires a complete ranking i.e. the balance between the positive and the negative outranking flows. The higher the net flow (Φ), the 'better' the option. A complete ranking can be obtained by using PROMETHEE II (**Figure D2.23**). In this case all the options are comparable. However, although no incomparabilities remain, the resulting information is more disputable because a considerable part of the information gets lost by considering the difference. It can be seen that only the first three options have a positive net flow.

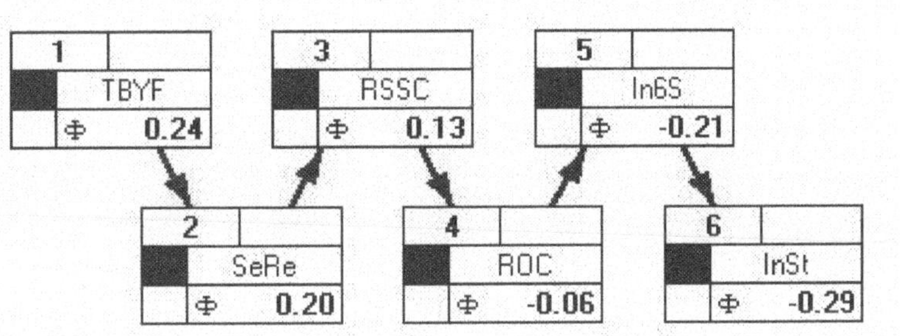

Figure D2.23: Complete Ranking (PROMETHEE II)

The Geometrical Analysis for Interactive Aid (GAIA) Plane

The GAIA procedure consists of a visual interaction module complementary to PROMETHEE. The GAIA plane provides clear graphical information about the conflicting character of the criteria and about the impact of the weightings on the final decision. The PROMETHEE I and II analyses are rather prescriptive, whilst GAIA is more descriptive and graphically oriented. **Figure D2.24** presents the GAIA plane for the analysis of the 6 options and 16 criteria. A magnified version is presented in **Figure D2.25**.

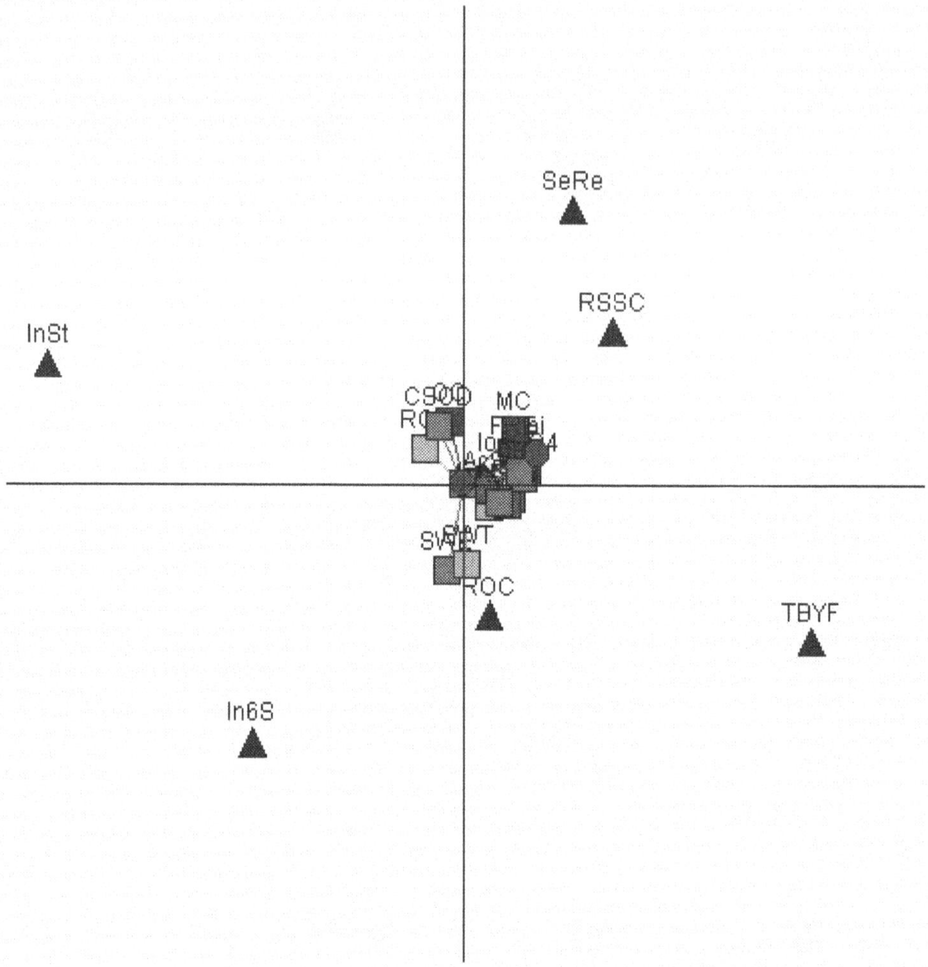

Figure D2.24: GAIA Plane

Figure D2.25: GAIA Plane (magnified)

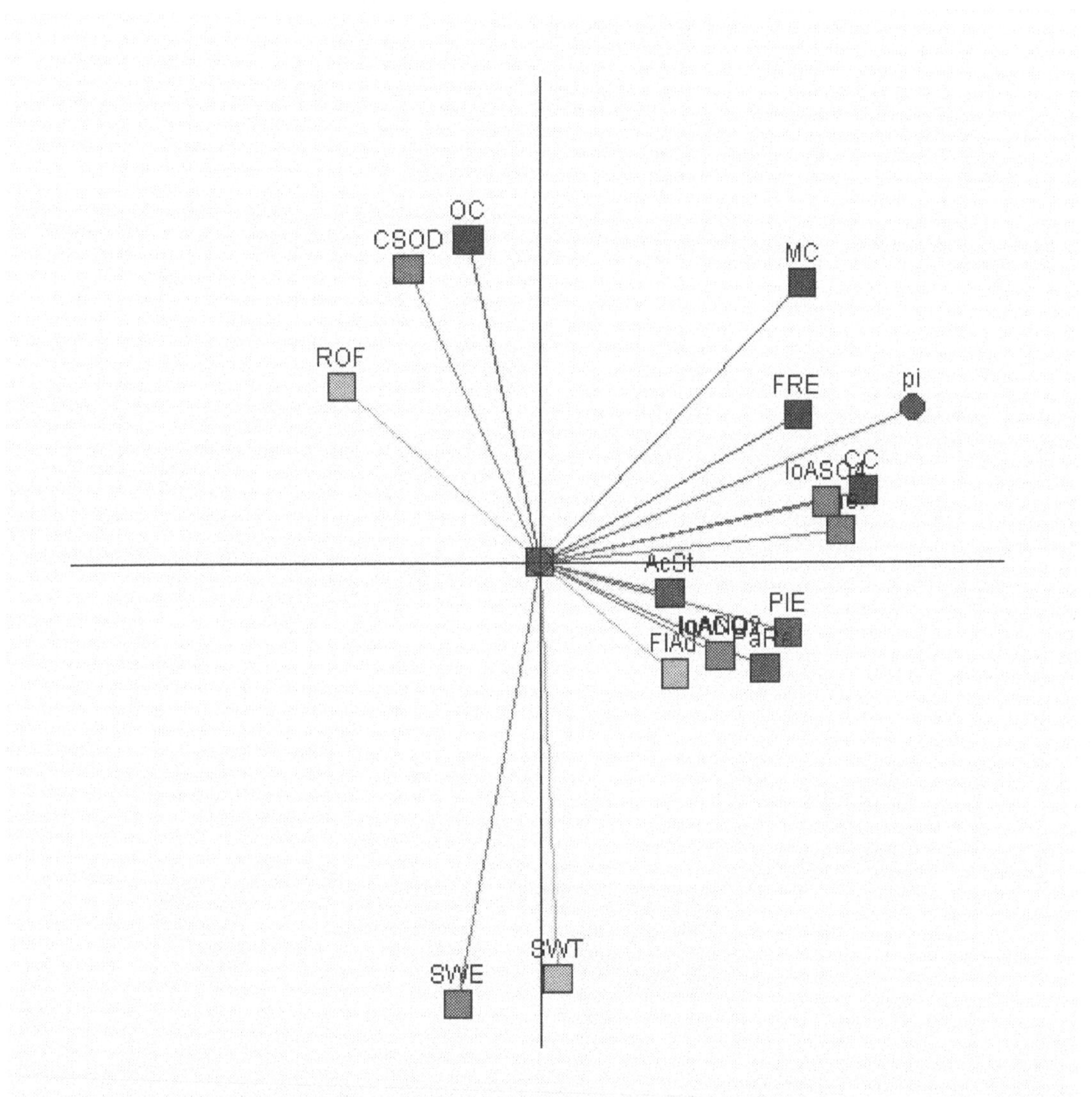

Figure D2.24 and D2.25 illustrate the following properties:

1. The longer a criterion axis in the GAIA plane, the more discriminating this criterion will be on the options (e.g. SWE, SWT).
2. Criteria expressing similar preferences on the set of alternatives will be represented in the GAIA plane by axes approximately oriented in the same direction (e.g. MC, FRE).
3. Criteria expressing conflicting preferences will be represented by axes in opposite directions (e.g. MC, SWE).
4. Independent criteria will be represented by orthogonal axes (e.g. OC, IoAC).
5. Options that are 'good' on a particular criterion will be represented by points located in the direction of the axis of this criterion (e.g. In6S, SWE).
6. Similar options will be represented by points located close to each other (e.g. SeRe, RSSC).

For modified weights, the positions of the criteria and the options/alternatives in the GAIA plane remain unchanged. The weight vector therefore looks like a decision stick that the decision-maker can move according to his preferences in favour of particular criteria. When the weights are changed, the stick (along with the PROMETHEE decision axis) moves so that the consequence for decision making can be easily observed. In this particular case the decision axis (the projection of the decision stick) is long and has a strong decision power. This indicates that the decision-maker is invited to select an option as far as possible in its direction (i.e. SeRe as shown in **Figure D2.24**).

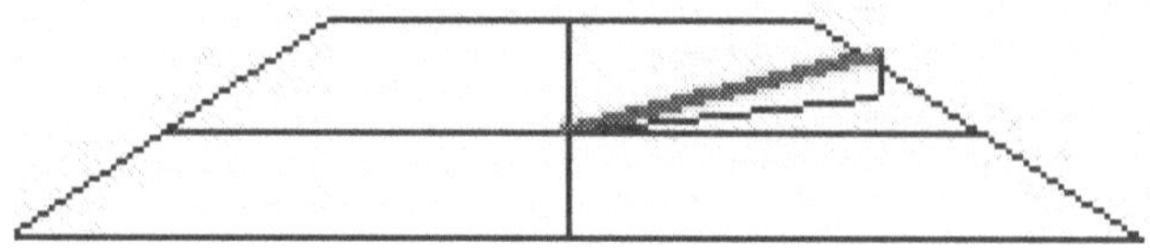

Figure D2.26: Decision Stick

Actions Profiles

Another positive feature of the PROMCALC software is the possibility of showing each option/action profile. This graphical illustration helps the decision-maker to notice the strengths and weaknesses of each option in terms of the selected criteria. The criteria on which the option is 'good' will be represented above the horizontal axes and vice-versa. The six actions/options profiles are presented **in Figures D2.27 to Figure D2.32.**

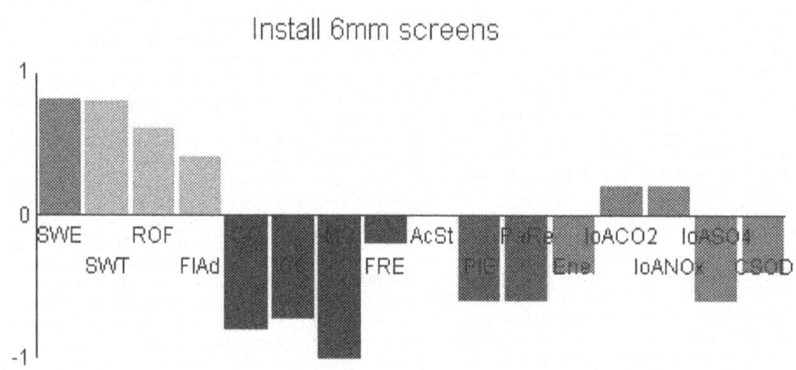

Figure D2.27: Install 6mm Screens Action Profile

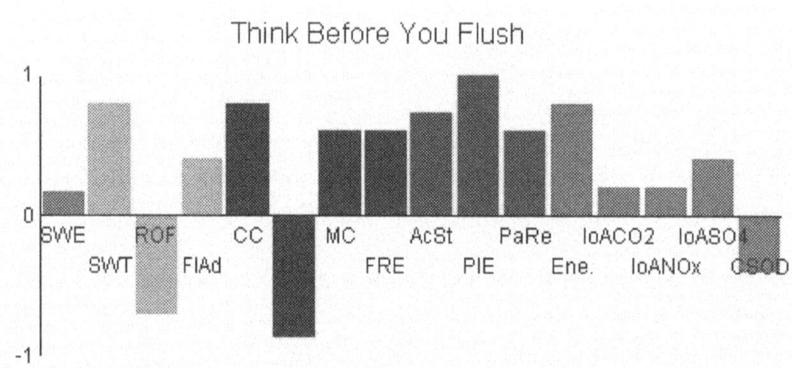

Figure D2.28: Think Before You Flush Action Profile

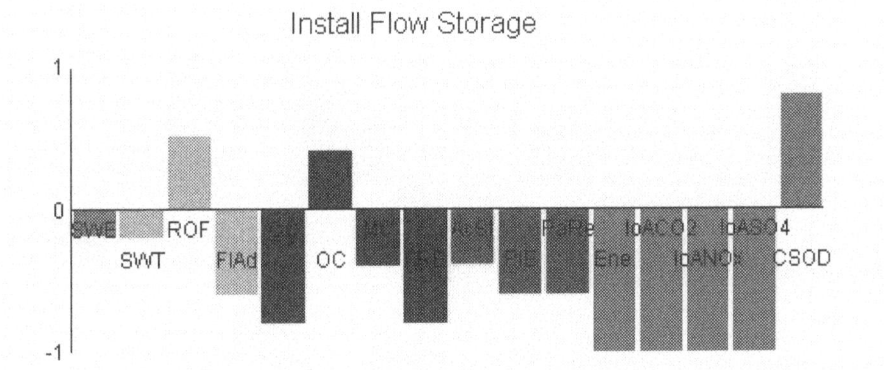

Figure D2.29: Install Flow Storage Action Profile

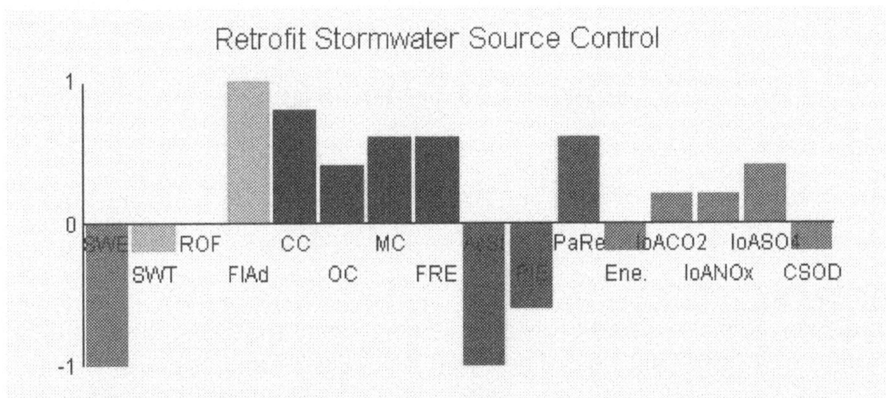

Figure D2.30: Retrofit Stormwater Source Control Action Profile

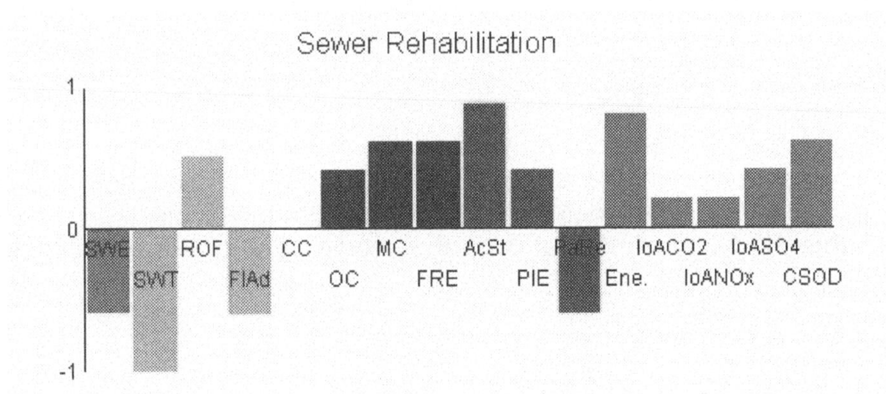

Figure D2.31: Sewer Rehabilitation Action Profile

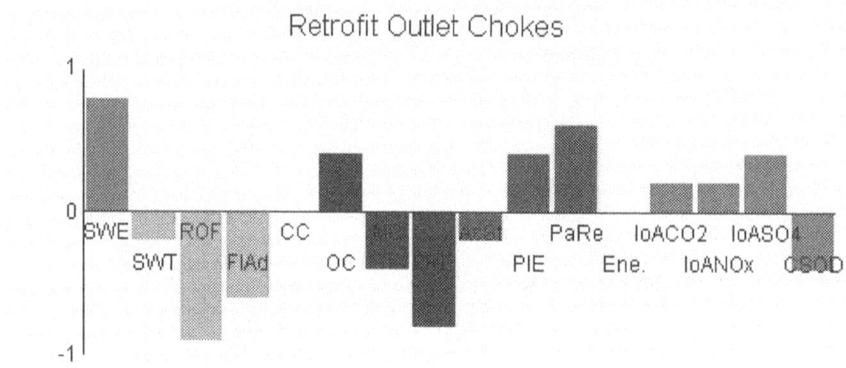

Figure D2.32: Retrofit Outlet Chokes Action Profile

The final ranking obtained using the PROMETHEE methods is presented **in Figure D2.33.** It indicates the preferred option to be TBYF followed closely by SeRe (sewer rehabilitation).

Figure D2.33: Final Ranking

TBYF 0.24 → SeRe 0.20 → RSSC 0.13 → ROC -0.06 → In6S -0.21 → InSt -0.29

ELECTRE Analysis

During the meeting with the decision-maker it was clear that for some criteria he found it difficult to choose one of the six predefined PROMETHEE functions. For this reason and the reasons explained previously the use of the ELECTRE III model was suggested. The ranking of the six options, in a decreasing order of preference, is based on a fuzzy outranking relation represented by a credibility degree. The value functions developed for the 16 criteria by members of the SWARD team used both the partial and complete pre-order using a descending distillation procedure by which the 'best' options are selected first each time from the remaining ones. The input data and parameter settings can be seen in **Tables D2.31 and D2.32**.

For each pseudo-criterion the concordance and discordance index was calculated using a pairwise comparison between the six options. Using the concordance and discordance indices, the credibility matrix was determined. The concordance indices are presented in the matrixes in **Table D2.33** and the matrix of the credibility degrees is presented in **Table D2.34**.

Table D2.31: ELECTRE input data

	SWE	SWT	RoF	FlAd	CC	OC	MC	FRE	AcSt	PIE	PaRe	EnUs	IoAC	IoAN	IoAS	CSOD
In6S	5.38	0	0	0.6	68000	628.67	3167	0.3	68.8	60.9	0	260	16400	18.9	423	58564
TBYF	36.86	0	0.75	0.6	10000	800	0	0	85.9	78.1	1	16.7	806	0	3.75	58564
InSt	52.70	0.5	0.15	0	70000	0	680	1	62.5	59.4	0	1410	150000	1880	1890	28062
RSSC	89.21	0.5	0.4	1	3925	0	0	0	51.6	57.8	1	216	7230	3.63	44.4	48480
SeRe	70.30	1	0.15	0	39000	0	0	0	89.1	71.9	0	39.7	3130	7.94	15	38607
ROC	6.54	0.5	1	0	32446	0	800	1	65.6	68.8	1	148	4590	31.1	60	56626

Table D2.32: ELECTRE parameter settings

	SWE	SWT	RoF	FlAd	CC	OC	MC	FRE	AcSt	PIE	PaRe	EnUs	IoAC	IoAN	IoAS	CSOD
Weight	8	4	9	7	10.5	10.5	10.5	3.5	4.5	7.5	3	3.5	3.5	3.5	3.5	8
Dir. of Pref.	Decreas	Decreas	Decreas	Increas	Decreas	Decreas	Decreas	Decreas	Increas	Increas	Increas	Decreas	Decreas	Decreas	Decreas	Decreas
Mode of Def.	Direct	Direct	Direct	Direct	Inverse	Direct	Direct	Direct	Direct	Direct	Direct	Direct	Direct	Direct	Direct	Direct
Indiff. Coef α	0	0	0	0	0	0	0	0	0	0	0	0	0	0	0	0
β	1	0.3	0.2	0.39	10000	1	200	0.2	1	4	0.3	99	50000	300	250	12000
Pref. Coef α	0	0	0	0	0	0	0	0	0	0	0	0	0	0	0	0
β	19	2	0.3	2	20000	500	500	2	10	5	2	100	50000	300	250	15000
Veto Coef α	-	-	-	-	-	-	-	-	-	-	-	-	-	-	-	-
β	-	-	-	-	-	-	-	-	-	-	-	-	-	-	-	-

Table D2.33: Concordance Matrix

	In6S	TBYF	InSt	RSSC	SeRe	ROC
In6S	**1**	0.59	0.71	0.64	0.41	0.53
TBYF	0.79	1	0.72	0.80	0.71	0.81
InSt	0.73	0.42	1	0.60	0.51	0.58
RSSC	0.78	0.76	0.75	1	0.72	0.80
SeRe	0.89	0.70	0.92	0.85	1	0.90
ROC	0.87	0.56	0.83	0.66	0.63	1

Table D2.34: Credibility Matrix

	In6S	TBYF	InSt	RSSC	SeRe	ROC
In6S	1	0.59	0.71	0.64	0.41	0.53
TBYF	0.79	1	0.73	0.80	0.71	0.81
InSt	0.73	0.42	1	0.60	0.51	0.58
RSSC	0.78	0.76	0.75	1	0.72	0.80
SeRe	0.89	0.70	0.92	0.85	1	0.90
ROC	0.87	0.56	0.83	0.66	0.63	1

The final ranking based on the previous calculation is presented in **Table D2.35**, where I indicates indifference and P preference. The * next to P indicates that the respective option is better placed than another in one of the outranking relations, the opposite being true for all the other outranking relations. R indicates that it is not possible to distinguish between the option ROC and the option RSSC at this point of the analysis.

Table D2.35: Ranking Matrix

	In6S	TBYF	InSt	RSSC	SeRe	ROC
In6S	I	P*	P*	P*	P*	P*
TBYF	P	I	P	P	P*	P
InSt	P	P*	I	P*	P*	P*
RSSC	P	P*	P	I	P*	R
SeRe	P	P	P	P	I	P
ROC	P	P*	P	R	P*	I

The outranking relationship between the six options is presented in a graphical form in **Figure D2.34** using the descending order (from the best option to the worst option) and in **Figure D2.35** using the ascending order (from the worst option to the best option). These two ways of ranking are called "distillations", and it can be seen that they are slightly different.

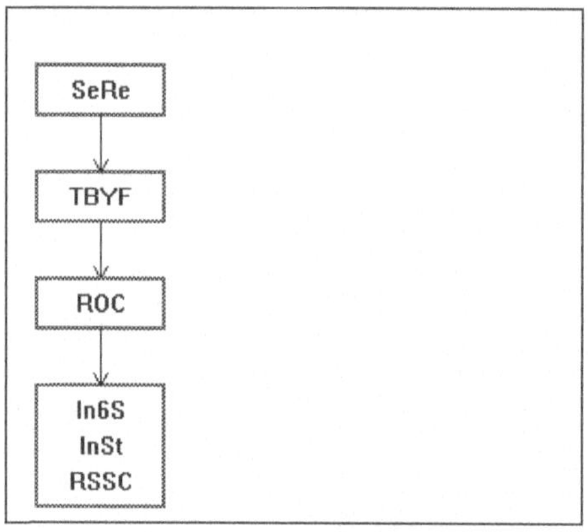

Figure D2.34 Descending distillation

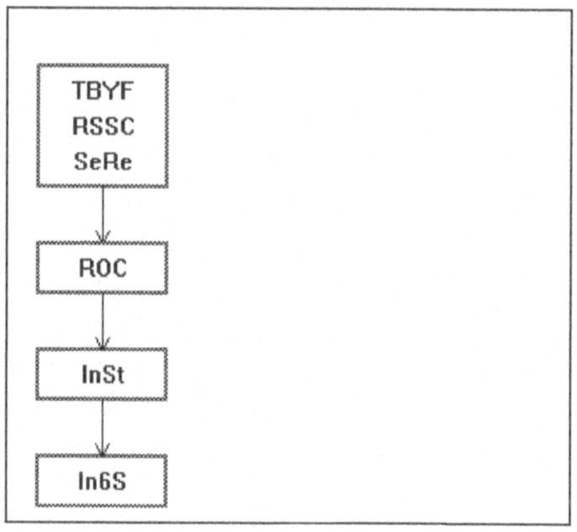

Figure D2.35: scending distillation

The median preorder is obtained by ranking the options (from the best to the worst) using the sum of the ranks obtained from the ascending and descending distillations. The median preorder is presented in **Table D2.36**.

Table D2.36: Median Preorder

Rank	Alternative
1	SeRe
2	TBYF
3	ROC
4	RSSC
5	InSt
6	In6S

The final ranking is obtained through the average of the ranks of the two intermediate (ascending and descending) distillations and is presented in **Table D2.37**.

Table D2.37: Ranks in Final Preorder

Rank	Alternative
1	SeRe
2	TBYF
3	RSSC ROC
4	InSt
5	In6S

The final ranking of the 6 options using ELECTRE III is presented in a graphical form in **Figure D2.36**. RSSC and ROC can be seen to have the same rank.

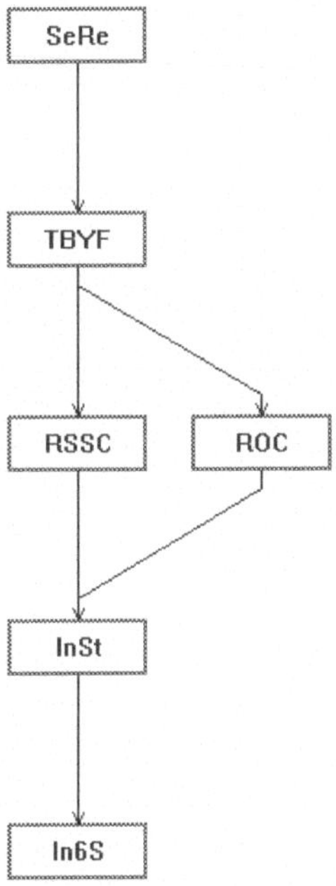

Figure D2.36: Final Graph

SMART Analysis

The SMART analysis can be seen in **Table D2.38** with a summary of the results in **Figure D2.37**.

Table D2.38: SMART Analysis

Criteria	Weight (Norm'd)	In6S Perf. Score	In6S Score	TBYF Perf. Score	TBYF Score	InSt Perf. Score	InSt Score	RSSC Perf. Score	RSSC Score	SeRe Perf. Score	SeRe Score	ROC Perf. Score	ROC Score
SWE	8	95	760	75	600	60	480	40	320	50	400	90	720
SWT	4	100	400	100	400	50	200	50	200	0	0	50	200
RoF	9	100	900	25	225	85	765	60	540	85	765	0	0
FlAd	7	60	420	60	420	0	0	100	700	0	0	0	0
CC	10.5	45	472.5	80	840	40	420	95	997.5	65	682.5	70	735
OC	10.5	85	892.5	80	840	100	1050	100	1050	100	1050	100	1050
MC	10.5	35	367.5	100	1050	80	840	100	1050	100	1050	75	787.5
FRE	3.5	70	245	100	350	0	0	100	350	100	350	0	0
AcSt	4.5	69	310.5	86	387	63	283.5	52	234	89	400.5	66	297
PIE	7.5	61	457.5	78	585	59	442.5	58	435	72	540	67	502.5
PaRe	3	0	0	100	300	0	0	100	300	0	0	100	300
EnUs	3.5	25	87.5	95	332.5	0	0	30	105	90	315	50	175
IoAC	3.5	80	280	95	332.5	70	245	85	297.5	90	315	95	332.5
IoAN	3.5	85	297.5	100	350	20	70	95	332.5	90	315	70	245
IoAS	3.5	60	210	95	332.5	10	35	85	297.5	90	315	80	280
CSOD	8	10	80	10	80	50	400	25	200	30	240	15	120
Score	**(100)**		**61.805**		**74.245**		**52.31**		**74.09**		**67.38**		**57.445**

The order of preference as determined from the SMART analysis is:

Figure D2.37: SMART ranking

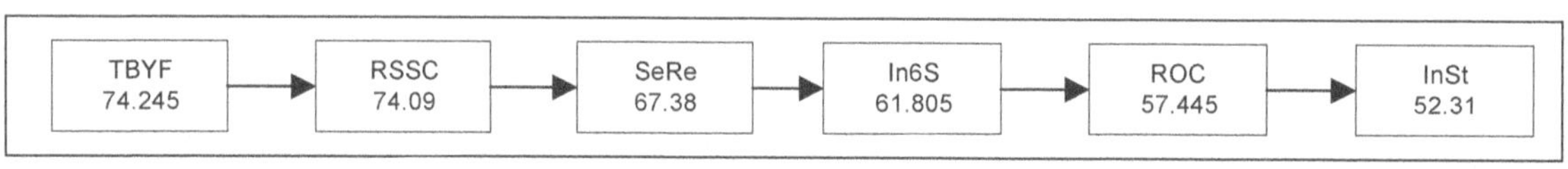

Comparison of results

The ranking of the options suggested by each of the methods of analysis is presented in **Table D2.39**.

Table D2.39: Summary of rankings

	PROMETHEE	ELECTRE	SMART
In6S	5th	6th	4th
TBYF	1st	2nd	1st
InSt	6th	5th	6th
RSSC	3rd	=3rd	2nd
SeRe	2nd	1st	3rd
ROC	4th	=3rd	5th

Overall consideration of the results would suggest that the TBYF option generally dominates the top rankings, followed by SeRe (sewer rehabilitation) and then the RSSC (retrofit source control). The other three options are quite consistently placed in the final three positions. However, whilst there would appear to be a reasonable degree of consistency between the outranking methods (PROMETHEE and ELECTRE), there is less consistency between these methods and the SMART analysis. This is likely to be due to the influence of trade-offs in the smart analysis between positive and less positive scores for individual criteria within the analysis.

D2.7.2 Alternative methods of data analysis

Reference Sustainability System (RSS)

Methodology for comparing options for sanitary waste disposal

Using the data collected, and the results of the detailed process modelling (using InfoWorks), a Reference Sustainability System (RSS) model was produced for the case study. The RSS model was developed and run in the form of an Excel spreadsheet.

Flows of both water and sanitary waste are represented in the model. These flows are linked using the methodology described in section D2.6.4.1 in which the working duration and proportion of flow to the storm outfall for each option were determined. From these figures and the total sanitary solids entering the system, the amount of sanitary solids escaping into the environment during storm incidents can be calculated for each option.

One of the advantages of using an RSS model as a complement to more detailed process modelling is that it enables the assessment of combinations of the basic options. In order to do this, the calculation of sanitary waste escaping was reproduced based on a function of the key parameter for each option. These were:

1. Screens at overflow — % solids captured at screen
2. TBYF campaign — % reduction of sanitary waste entering the system
3. Flow storage — % reduction of stormwater
4. Source control — % reduction of stormwater
5. Sewer rehabilitation — % reduction of stormwater
6. Retrofit constricting WCs — % reduction of sanitary waste entering system.

For options 4 and 5, for which a number of process model runs had been done, it was possible to interpolate between the particular cases to estimate the storm flow and sanitary waste escaping for any value of the key parameter. For the other options, a linear interpolation was applied. It was then possible to further estimate the storm flow and sanitary waste escaping for a combination of options by applying a linear combination of the parameters. It is important to note that, because the output (sanitary waste escaping) for each option is not a linear function of the input parameter, applying linear combinations of parameters does not produce a linear combination of the outputs. Thus, whilst this approach does introduce further approximation, it should produce a more accurate estimate for the sanitary waste escaping for the case of a combination of options than would be given by simply adding up the outputs for the individual cases.

The spreadsheet format of the RSS model also makes it suitable for calculating and comparing the cost effectiveness of options and combinations of options. This was done by applying the standard economic technique of calculating the levellised cost per unit of sanitary waste reduction for each option. This was calculated based on a 20-year life span for each option, as follows:

- For each option, the capital cost in the initial year and the revenue (operational and maintenance) cost per year were calculated using the data described in section D2.6.1
- The sanitary waste reduction for each year was calculated using the RSS model with the appropriate value of the parameter for that option.
- For each option, the 'net present value' of both the costs and the sanitary waste reduction were calculated over the 20-year period, using a 6% discount rate.
- The levellised cost per unit of sanitary waste reduced (in £/kg) is then calculated by dividing the net present cost by the net present value of the sanitary waste reduction. The value of the levellised cost for each option is shown in **Table D2.40** and **Figure D2.38**. (For each of options 4 and 5, the central case was chosen, i.e. options 4-b and 5-c.)

Table D2.40: Sanitary waste reduction and cost per unit reduction for options 1 to 6

Option	Sanitary waste reduction (kg/year)	Running , Repair and Maintenance cost (£)	Capital cost (£)	Cost per unit reduction (£/kg)
1	102.3	3,796	68,000	92
2	70.8	800	10,000	23
3	55.0	680	70,000	117
4_b	16.8	0	3,925	19
5_c	37.4	0	39,000	86
6	101.2	800	32,446	34

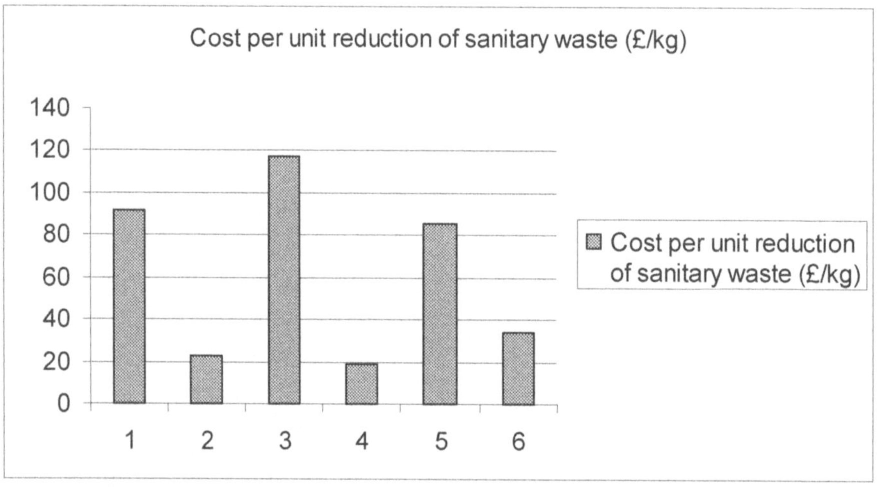

Figure D2.38: Cost per unit reduction of sanitary waste for options 1 to 6

On the basis of these levellised costs, it is then possible to construct an 'abatement graph', which represents the most cost-effective way of meeting a desired abatement target. In this case, the abatement target is the reduction of sanitary waste escaping into the environment to a level equivalent to installing a 6 mm screen on the overflow (Option 1). As the screen is taken to be 95% efficient, this represents a reduction by 102.3 kg/year.

The most cost-effective option is Option 4-b, retrofitting stormwater source control with 35% of roof area controlled, at a cost of £19/kg reduced. Carrying out this option reduces the annual sanitary waste discharge by 16.8 kg/year. The second most cost-effective option is Option 2, carrying out a 'Think Before You Flush' (TBYF) campaign. To ascertain the combined effect of Options 4 and 2, the RSS model was run with the parameters for both options set, i.e. a 65.8% reduction in sanitary solids and a 35% reduction in stormwater. This was found to result in a reduction by 76.6 kg/year for the combined case, at a cost of £27/kg for the additional reduction.

Note that, as described above, this is a smaller reduction than would be anticipated from adding the individual reductions, reflecting the non-linearity of the calculation. This is also reflected in the higher unit cost for the additional saving, as it becomes more expensive to make further reductions.

The third most cost-effective option is Option 6, retrofitting constricting WCs. Combining Options 4, 2 and 6 results in a cumulative reduction of 105.8 kg/year, thus achieving the abatement target of matching the effectiveness of 6 mm screens. The additional cost of Option 6 in this case is £119/kg, again reflecting the high cost of making further reductions. However, the average levellised cost of undertaking Options 4, 2 and 6 is £51/kg. This means that this combination of Options is more cost-effective than Option 1, which achieves a similar reduction at a cost of £86/kg.

The abatement graphs for the two alternatives are shown **in Figures D2.39 and D2.40.**

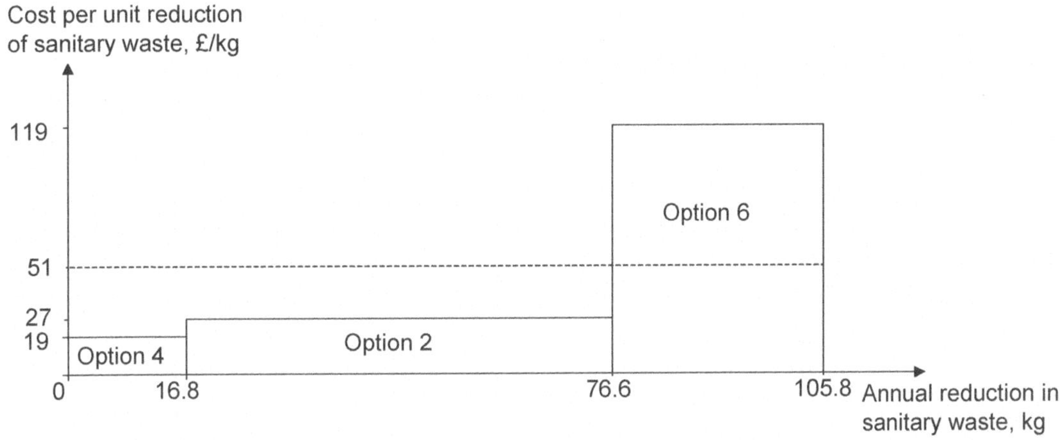

Figure 39: Abatement graph for combination of Options 4, 2 and 6

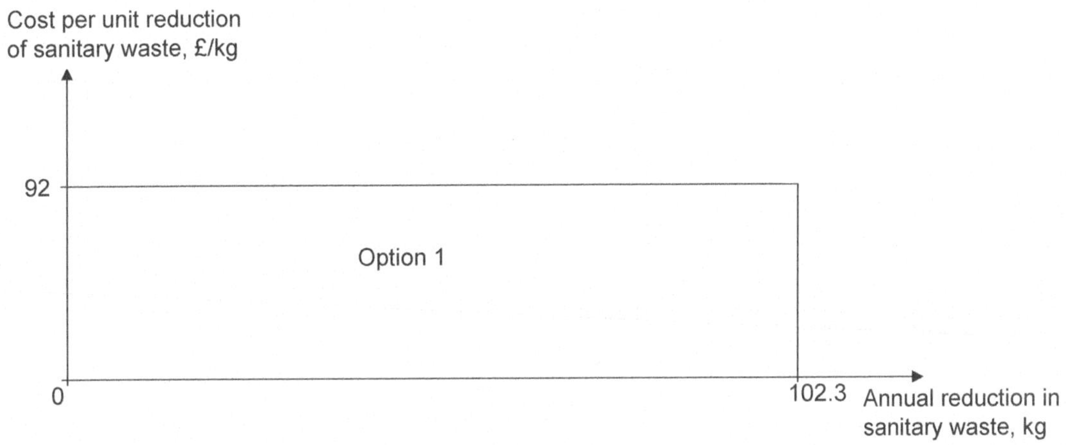

Figure 40: Abatement graph for Option 1

Discussion of the RSS

The above analysis has demonstrated how a systems model, such as the RSS, can complement the use of more detailed process modelling. In this example, the use of a systems model enabled the cost effectiveness and waste reduction resulting from combinations of options to be assessed. A particular combination of options was found to be a more cost-effective way of achieving the required abatement target than the traditional option of installing screens on the overflow pipe. This combination of options could be tested again using the detailed process model, in order to verify the waste reduction estimated using the systems model approximation.

However, a number of points need to be made regarding any interpretation of this combination as the 'optimum' solution to be problem of finding the most sustainable way of reducing sanitary waste emissions to the environment.

Firstly, as the systems model relies on approximations, it is most useful as a tool for looking at the 'big picture' in order to see how system elements interact, whilst potentially losing information about the details of the system. As for any modelling tool, systems models rely on the quality of data obtained. In this case, most of the relevant data was collected for the detailed process modelling stage, which made the construction of the systems model easier. In general, the additional benefits obtained from the use of a systems model need to be weighed against the extra time and effort needed for its construction.

A more serious limitation was that, in this case, the systems model only incorporated a limited number of criteria - namely, the economic criteria of capital and operating and maintenance costs, and the environmental criterion of sanitary waste reduction. As the multi-criteria analysis methodologies showed, the assessment of the most sustainable option depended on taking into account the full range of economic, environmental, social and technical criteria. It would be possible to incorporate further quantitative criteria into the RSS or other systems model, at the cost of making the model significantly more complex. In addition, a further level of multi-criteria analysis would then be necessary to assess the most sustainable option.

However, it is possible to combine qualitative criteria with the systems modelling. For example, though the combination of Options 4, 2 and 6 is found to be more cost-effective in reducing sanitary waste emissions than Option 1, a water company would also want to compare these cases using the criteria of risk of failure, flexibility and adaptability, and participation and responsibility. Both Option 2 (TBYF campaign) and Option 6 (WC retrofit) have high or very high risk of failure in terms of meeting sanitary waste objectives, meaning that the cost-effective combination compares unfavourably with the very low risk Option 1 (screens) on this criterion. Similarly, the combination including Option 6 would have a lower adaptability than Option 1, though Options 2 and 6 engender higher levels of participation.

This case study has demonstrated the use of systems modelling, such as the Reference Sustainability System approach, to complement the use of more detailed process modelling. In particular, the use of a systems model enables relatively easy assessment of combinations of the basic options and comparison of the cost-effectiveness of options and combinations of options in meeting particular abatement targets. A conceptual representation of the RSS systems model is shown **in Figure D2.41.**

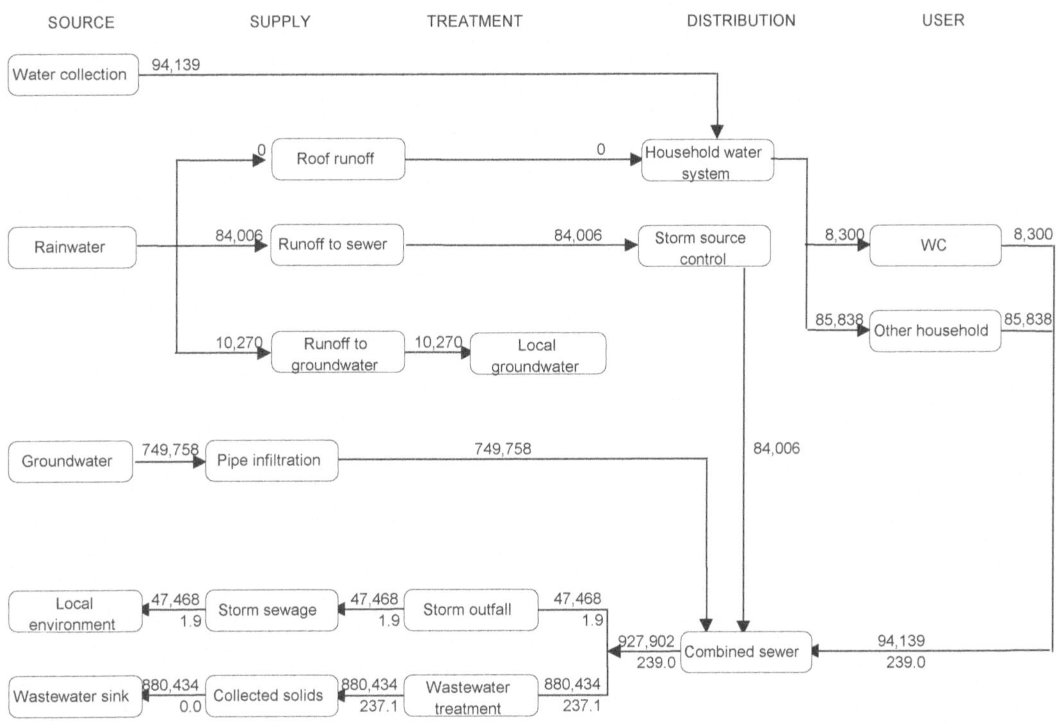

Key:
Water flow (m^3/year)
Solid flow (tonnes/year)

Figure D2.41: Conceptual representation of RSS systems model

EcoIndicator
The Eco-indicator approach that can be utilised by SimaPro within the LCA was briefly discussed in section D2.6.2. The approach could be used as an alternative method to gauge the relative sustainability of a range of options, although this is obviously limited as its assessment is only of environmental impacts. The Eco-Indicator 95 uses the distance to target approach to analyse impacts, including damage to both human health and to ecosystems. Weightings are established by assessing the current and target values for an effect. The greater the distance the more serious the effect. Extensive analysis of weighting methods and of around one hundred LCAs was undertaken by Goedkoop (1995) in order to calculate the indicators.

The more complex Eco-indicator 99 uses a damage function approach rather than a distance to target principle. The damage function approach is a more robust methodology with a sounder scientific basis, which relates the impact to the damage to human health or to the ecosystem. A weighting system between the damage categories has been developed through consultation based on individual weightings and perspectives. The controversial weighting stage in LCA has been simplified by this method.

Life cycle results
The characterisation results from the SimaPro model using the Eco-indicator 95 are shown in **Figure D2.42**. Due to the potential recycling of the stainless steel screen and concrete storage tank there is a positive impact for the environment, offsetting the negative impacts. The option 2 results (TBYF campaign) show the lowest environmental impact for all calculated parameters except heavy metals and summer smog, which are high as a result of

the processes involved in paper manufacturing. The normalised effects of an average European inhabitant over one year are shown in **Figure D2.43**, with **Figure D2.44** showing the evaluation values. In the ecopoints method the distance-to-target principle is used in the calculation of these values. In the evaluation phase the normalised effects scores are multiplied by a weighting factor representing the relative importance of the effect. The height of the column in **Figure D2.44** represents the seriousness of the effects on human health and ecosystems. The seriousness is measured in indicator points.

Figure D2.45 shows the results of the evaluation method in a single figure for the environmental impacts. SimaPro can add together all the different environmental effects from the evaluation stage to give a total impact for each material and process in the life cycle. The value mode in SimaPro displays the effect scores in tabular form. The graph shows that option 1 (6 mm screen) has the highest value for environmental effects whilst option 2 (TBYF campaign) shows the lowest values.

The Eco-indicator method allows the major relative environmental impacts of the proposed options to be identified. The greatest impacts occur at the assembly/raw material stage. The waste disposal scenario for option 3 (storage tank) and option 1 (6 mm screen) offsets the total indicator value, as more than 70% of the screen and storage tank total weight is recycled. The energy results show the highest value amongst all the options is for the option 3 (the storage tank), followed by the option 4 (rainwater barrels). The analysis shows that the life cycle indicator values for option 4 (rainwater barrels) are higher than for the screen with regard to heavy metals, greenhouse gases, ozone and summer smog, but the total indicator value for the screen is higher. Overall, the analysis shows that option 2 (the TBYF campaign) has the least environmental impact.

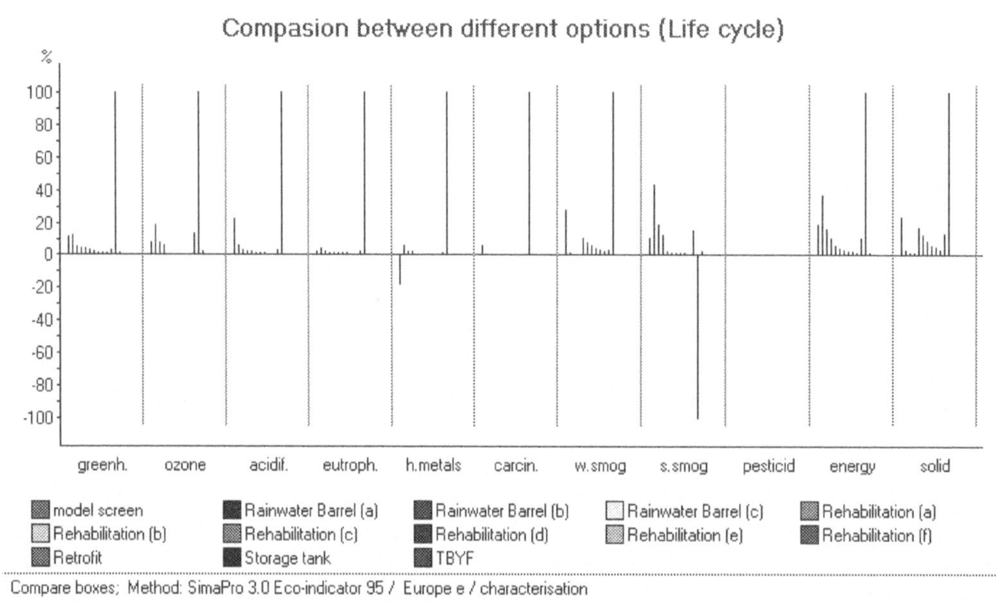

Figure D2.42: Characterisation from Eco-indicator 95

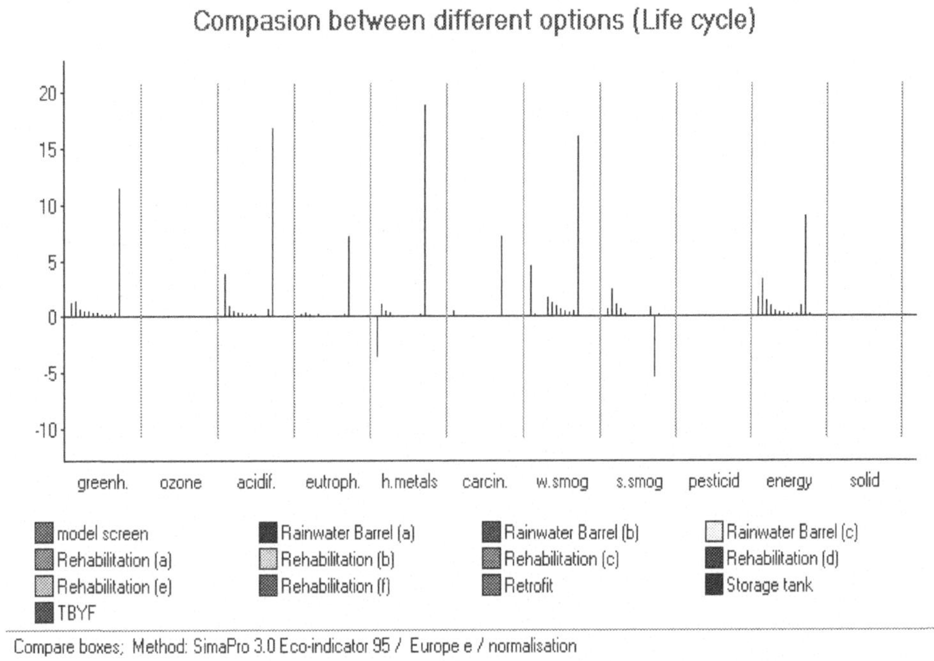

Figure D2.43: Normalisation (per inhabitant) for Eco-indicator 95

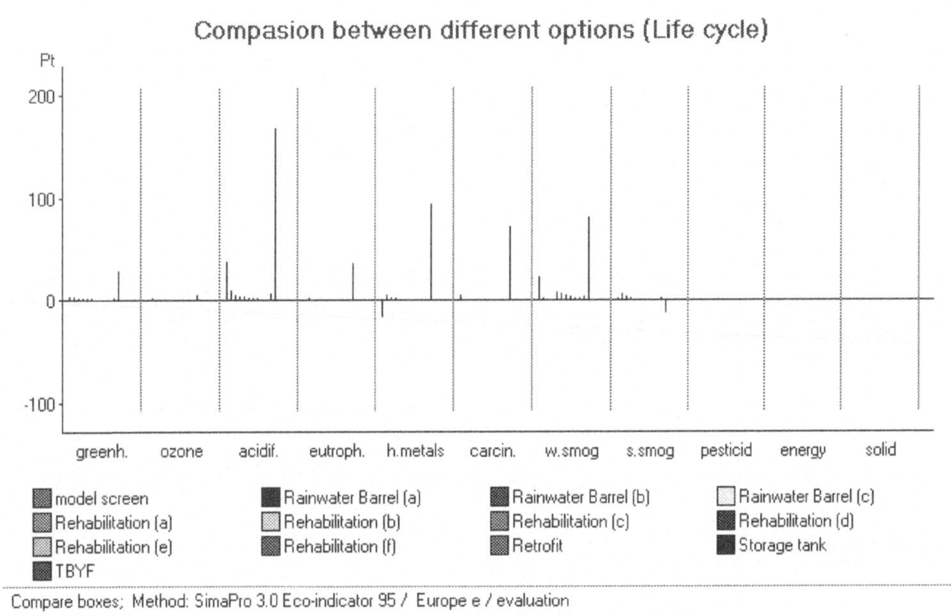

Figure D2.44: Evaluation ecopoints

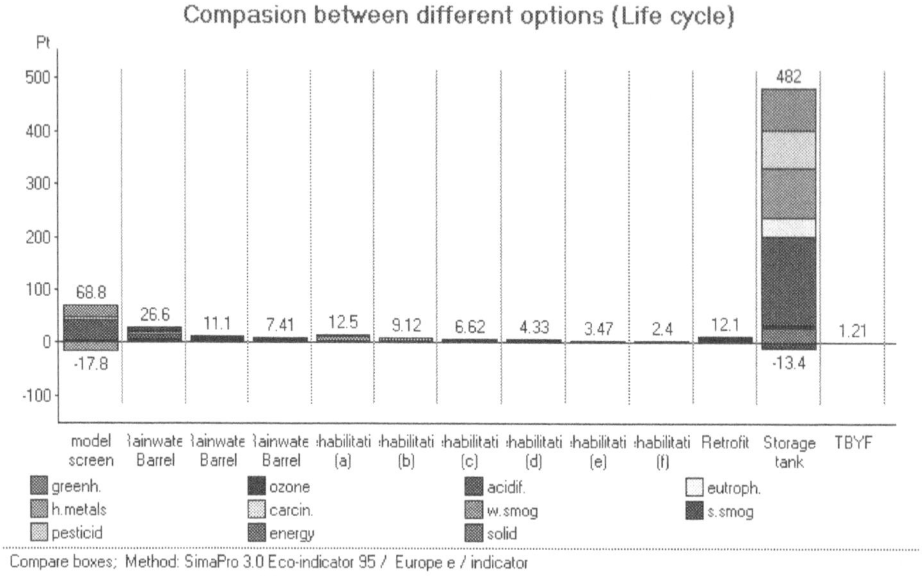

Figure D2.45: Unified Indicators

LCA Sensitivity Analysis
The waste scenario aspect of the LCA was chosen as the focus for LCA sensitivity analysis as this was the aspect of the chain for which there was the least information available. A change in the percentage of different disposal scenarios was assumed then the percentage of change in air emission and energy indicator were calculated.

Option 1 (6 mm screen)
Table D2.41 present the sensitivity analysis results of this option. The disposal scenario used in the original analysis was 30% landfill, 60% recycle and 10% reuse.

Table D2.41: Results of variations in disposal scenario – option 1 (6 mm screen)

Scenario No	Disposal scenario (%)			Indicator Change (%)				
	Landfill	Recycle	Reuse	CO_2	SO_4	NOx	SO_2	Energy
1	90	0	10	4	1	11	1	22
2	60	30	10	1	1	5	1	11
3	45	45	10	0	0	3	0	6
4	30	60	10	0	0	0	0	0
5	0	90	10	-5	-1	-5	-1	-11

From the results of the analysis, it can be seen that the energy and air emission indicator decreases in relation to the percentage of material being recycled. The results imply that energy is the most sensitive indicator to the changes in waste scenarios. They also indicate that the emissions and energy used for the assembly/construction stage is the main contributor from the whole life cycle of the screen option.

Option 2 -Think Before You Flush (TBYF) campaign
Table D2.42 presents the sensitivity analysis results of the TBYF option. The disposal scenario used in the original analysis was 100% landfill.

Table D2.42: Results for variations in disposal scenario – Option 2 (TBYF)

Scenario No	Disposal scenario (%)			Indicator Change (%)				
	Landfill	Recycle	Reuse	CO_2	SO_4	NOx	SO_2	Energy
1	100	0	0	0	0	0	0	0
2	90	10	0	-1	-3	0	0	-7
3	60	40	0	-4	-15	0	0	-31
4	45	55	0	-2	-21	0	0	-42
5	0	100	0	-7	-39	0	0	-77

The results show that the indicators for energy and air emissions decrease in relation to the percentage of material being recycled. The results imply that paper recycling can recover energy (77%) as well as reduce SO_4 emissions by 39%. The results also show the energy as most sensitive indicator for changing in waste disposal scenario amongst all indicators.

Option 3 - Storage tank

Table D2.43 presents the results of the sensitivity analysis for option 3. The disposal scenario of this option in the original analysis was 30% landfill and 70% recycling.

Table D2.43: Results for variations in disposal scenario – Option 3 (storage tank)

Scenario No	Disposal scenario (%)			Indicator Change (%)				
	Landfill	Recycle	Reuse	CO_2	SO_4	NOx	SO_2	Energy
1	100	0	0	5	0	1	-3	82
2	80	20	0	3	0	1	-2	79
3	45	55	0	1	0	0	-1	18
4	30	70	0	0	0	0	0	0
5	0	100	0	-2	0	0	1	-35

The results of the sensitivity analysis show that energy is the only really sensitive indicator in regard to the different waste scenarios.

Option 4 – Source control

Table D2.44 presents the results of the sensitivity analysis on the source control option. The waste scenario used for the original analysis was 50% landfill and 50% recycle.

Table D2.44: Results for variations in disposal scenario – Option 4 (source control)

Scenario No	Disposal scenario (%)			Indicator Change (%)				
	Landfill	Recycle	Reuse	CO_2	SO_4	NOx	SO_2	Energy
1	0	50	50	21	-2	-315	353	-38
2	30	50	20	81	58	-260	412	22
3	50	50	0	0	0	0	0	0
4	50	30	20	73	58	-134	271	37
5	70	20	10	88	78	-53	220	65

From the results of the analysis, the energy and air emission indicators increases in relation to the percentage of material being landfill. The results show that the air emissions indicators are more sensitive than the energy indicator for the changing waste scenarios.

Option 5 – Sewer rehabilitation

Table D2.45 presents the results of the sensitivity analysis for the rehabilitation option. The original disposal scenario for this option was 50% landfill and 50% recycled.

Table D2.45: Results for variations in disposal scenario – Option 5 (sewer rehabilitation)

Scenario No	Disposal scenario (%)			Indicator Change (%)				
	Landfill	Recycle	Reuse	CO_2	SO_4	NOx	SO_2	Energy
1	0	100	0	-10	0	0	0	-11
2	25	75	0	-5	0	0	0	-5
3	50	50	0	0	0	0	0	0
4	75	25	0	5	0	0	0	5
5	100	0	0	10	0	0	0	10

The results of the sensitivity analysis for this option show that CO_2 and energy are the only indicators affected by the change in the waste scenarios. The sensitivity of the indicators to the variations in waste disposal scenario is relatively low.

Option 6 – Retrofit constricting WCs

Table D2.46 presents the results of the sensitivity analysis for the retrofit option. The original disposal scenario for this option was 92% landfill and 8% recycling.

Table D2.46: Results for variations in disposal scenario – Option 6 (retrofit WCs)

Scenario No	Disposal scenario (%)			Indicator Change (%)				
	Landfill	Recycle	Reuses	CO_2	SO_4	NOx	SO_2	Energy
1	92	8	0	0	0	0	0	0
2	80	20	0	2	-2	-5	29	-4
3	60	40	0	4	-5	-14	78	-11
4	45	55	0	6	-7	-20	114	-16
5	0	100	0	11	-13	-39	225	-32

The results show that the energy indicator decreases in relation to the percentage of material being recycled. The results imply that SO_2 emissions are more sensitive than the energy indicator for the change in waste scenarios, and increases with an increase in the amount of material being recycled.

The results from the sensitivity analysis of all six options indicate that the air-emission indicators for option 4 (source control) are the most sensitive to the changes in disposal scenarios. The second most sensitive option is Option 6 (retrofitting of constricting WCs). The energy indicator appears to be the most sensitive to these changes for the other options (1, 2, 3 and 5). Despite the changes in disposal scenarios used in the sensitivity analysis, Option 3 (storage tank) still has the highest air emissions and energy use, and thus the greatest impact on air quality. The TBYF campaign remains the option with the least impact on air.

D2.8 OPTION SELECTION

The option selection stage involves a risk assessment on the analysis described in Section D2.7 above and the application of human judgement and experience:

- to facilitate the inclusion of any intangible aspects not yet considered; and
- to reach a final decision.

D2.8.1 Risk Assessment

Risk Identification

In the context of the decision making process, the risk of reaching a wrong conclusion arises from variability and inaccuracy in the weights and scales used in the analysis in phase 5.

Variability in the weights of group and individual criteria.

The weightings for group and individual criteria were developed by reaching a consensus between the participants at the workshop and are therefore considered to be robust. Nevertheless, sensitivity testing of the effects of changes to the weightings was undertaken as described below.

Variability in the scores for individual criteria.

Variability in the scores of the criteria was identified as follows:

- Criteria were based on the output of models - the estimated or known accuracy of the model were taken as an indication of accuracy of the scores;
- Economic Criteria – based on an estimate of the accuracy of cost estimates at the conceptual stage of design.
- Social Criteria – based on the Standard deviation of the individual responses as presented in Tables 16 and 17.

The variability represented in the simulation is shown in **Table D2.47**.

Table D2.47: Variability of criteria scores

Individual criteria	Variability
Capital cost	±30
Operational cost	±30
Maintenance	±30
Sanitary waste escape	±20
Primary energy	±30
Impact on air	±30
CSO discharge	±20
Social Criteria	See Standard Deviations Tables 16 and 17

Risk Analysis

Sensitivity Analyses

Sensitivity analyses were performed on each of the three methods by way of changing the value of each of the weights. In this case, the sensitivity analysis was carried out by inverting the weighting function of each of the criteria, and by equal weighting.

Inverse Weightings

Inverting the weightings were achieved in a similar fashion to the distribution of weights, but where both the collective prioritisation of the four groups and the inter-category importance of each of the grouped criteria were reversed. For example, using the same criteria as before (RoF), the Technical group's prioritisation was reversed to have 30% importance (from 20%), and the inter-category importance of the RoF criteria was reversed to have 20% importance (from 45%), thus giving an 'inverse' weighting of 0.3 × 0.2 = 0.06, and therefore a weighting factor of 6. In this context, the 'inverse' weightings of all the criteria are detailed in **Table D2.48**, and which were fed into the three decision-aiding analyses labelled as 'inverse' weights.

Table D2.48:Inverted weights for sensitivity testing

Criteria	'Inverse' Weighting
CC	2.5
OC	2.5
MC	2.5
FRE	7.5
AcSt	10.5
PIE	7
PaRe	17.5
SWE	2
EnUs	4
IoAC	4
IoAN	4
IoAS	4
CSOD	2
SWT	13.5
RoF	6
FlAd	10.5
Total	**100**

Equal Weightings

Equal weightings of 6.25 were assigned to each of the criteria (16 × 6.25 = 100). This was done to keep consistency with the resultant values of the analyses, especially with respect of the SMART analysis.

Results of Change in Weightings

The final option rankings for each of the methods is shown in **Table D2.49**.

Table D2.49: Options Rankings following sensitivity analysis

	PROMETHEE			ELECTRE			SMART		
	Normal	Inverse	Equal	Normal	Inverse	Equal	Normal	Inverse	Equal
In6S	5th	5th	5th	6th	=4th	5th	4th	3rd	4th
TBYF	1st	1st	1st	2nd	1st	=1st	1st	1st	1st
InSt	6th	6th	6th	5th	6th	6th	6th	6th	6th
RSSC	3rd	2nd	3rd	=3rd	3rd	=3rd	2nd	2nd	2nd
SeRe	2nd	3rd	2nd	1st	2nd	=1st	3rd	5th	3rd
ROC	4th	4th	4th	=3rd	=4th	=3rd	5th	4th	5th

In terms of sensitivity, despite the major changes made to the weightings, the rankings of the options identified by the three methods appears to be quite robust. For the PROMETHEE analysis, it can be seen that TBYF, representing the 'Think Before You Flush' campaign option, appear as 'top' rank in two of the three sets of weightings and appearing as third rank for the other. Similarly, the RSSC option appears as top rank for one set and second rank for the other two. The SeRe and In6S options generally share the third and fourth ranks, and the ROC and InSt are constant in the fifth and sixth ranks across all three weighting sets.

For the ELECTRE analysis, the SeRe option is the top rank across all three sets (with one shared), with the RSSC and TBYF options generally sharing the second and third ranks. The ROC and InSt options are generally placed in the fifth and sixth ranks again. The In6S option, however, appear to be a little sensitive to the weight changes.

The SMART analysis shows that the rankings are robust despite the varying weights. The TBYF and RSSC and InSt options are unchanged with some movement in position between the other options.

For the inverted weighting set, the TBYF option dominates the top ranking across all three analyses. The RSSC option is quite robust, appearing as second rank for two of the analyses, whilst the SeRe option appears in third rank for all three analyses. Also quite robust, is once again the ROC and InSt options in fifth and sixth places. Finally, for the equal weighting set, the top three places are generally constant with the TBYF, RSSC, and SeRe options in order of general preference. The In6S, ROC, and InSt are unwavering in the fourth, fifth, and sixth places that they occupy.

Further sensitivity testing was carried out on the SMART analysis and the results of this are shown **Table D2.50 and Figure D2.46**.

Table D2.50: Smart Analysis Sensitivity testing results

Options	Sensitivty Test Number								
	1	2	3	4	5	6	7		
SCREEN	62	59	61	63	60	62	62		
TBYF	74	86	72	76	74	74	74		
STORAGE	52	34	54	54	51	54	52		
SOURCE	74	76	70	75	71	75	74		
REHAB	67	50	66	69	65	68	67		
RETROFIT	57	56	55	60	55	58	57		

Key:

1 Initial Results
2 Weights Inverted
3 Weights equal
4 Score for capital cost +30%
5 Score for capital cost -30%
6 Score for CSO Discharge +30%
7 Score for CSO Discharge -30%

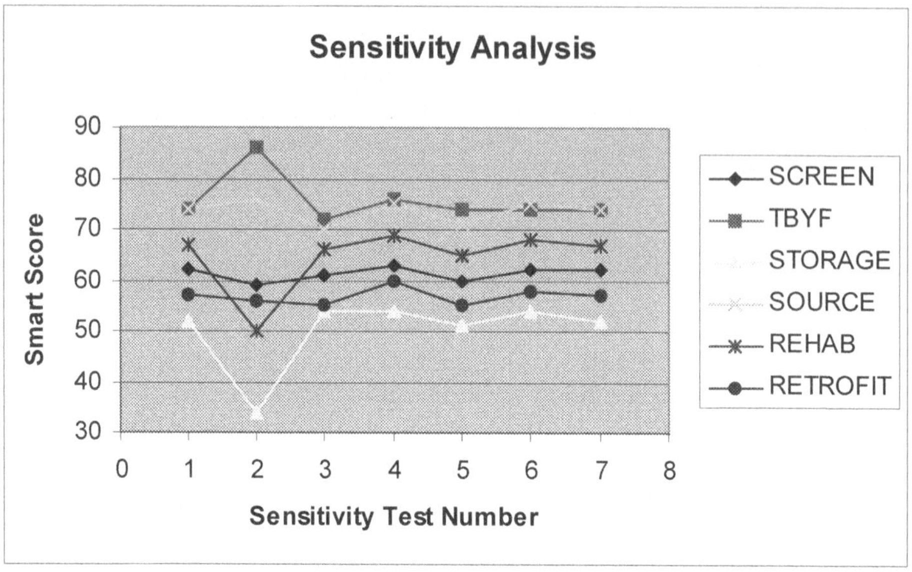

Figure D2.46: Smart Analysis Sensitivity testing results

Based on the overall results of the three models the final ranking can be considered as that shown in **Table D2.51**.

Table D2.51: Final Ranking

Rank	Alternative
1	TBYF
2	SeRe
3	RSSC
4	ROC
5	In6S
6	InSt

It is the decision-maker's responsibility to consider if in selecting the TBYF option, the assumptions made about public changes in culture and habit are not too optimistic. The final decision depends very much on the attitude to risk the decision-maker has, i.e. risk-taker, neutral to risk or adverse to risk. Both 'investment' options - SeRe and RSSC - are very close in terms of their utility and can complement the TBYF option in the case of an adverse to risk or neutral to risk decision-maker. The MCA ranking is in a stark contrast with the one established on the basis of professional judgement (with no MCA) by the decision-makers in a workshop. Their ranking is presented in **Table D2.52**.

Table D2.52: Decision-maker's ranking

Rank	Alternative
1	In6S
2	InSt
3	SeRe
4	RSSC
5	TBYF
6	ROC

Table D2.52 shows clearly that the decision-maker have top ranked the poorest options and emphasis the necessity to undertake a MCA especially for strategic decisions.

Monte Carlo simulation

Further risk assessment was undertaken on the SMART analysis using monte-carlo simulation with the scores for criteria being replaced by probability distributions. In the absences of sufficient data to allow the precise nature of the probability distributions to be determined, truncated normal distributions were used. Truncation was applied to prevent scores less than 0 or greater than 100 to be used and the standard deviation of the distribution were selected on the basis of the variability on scores described in section D2.8.1 above.

The results of the simulation are shown in **Figure D2.47**.

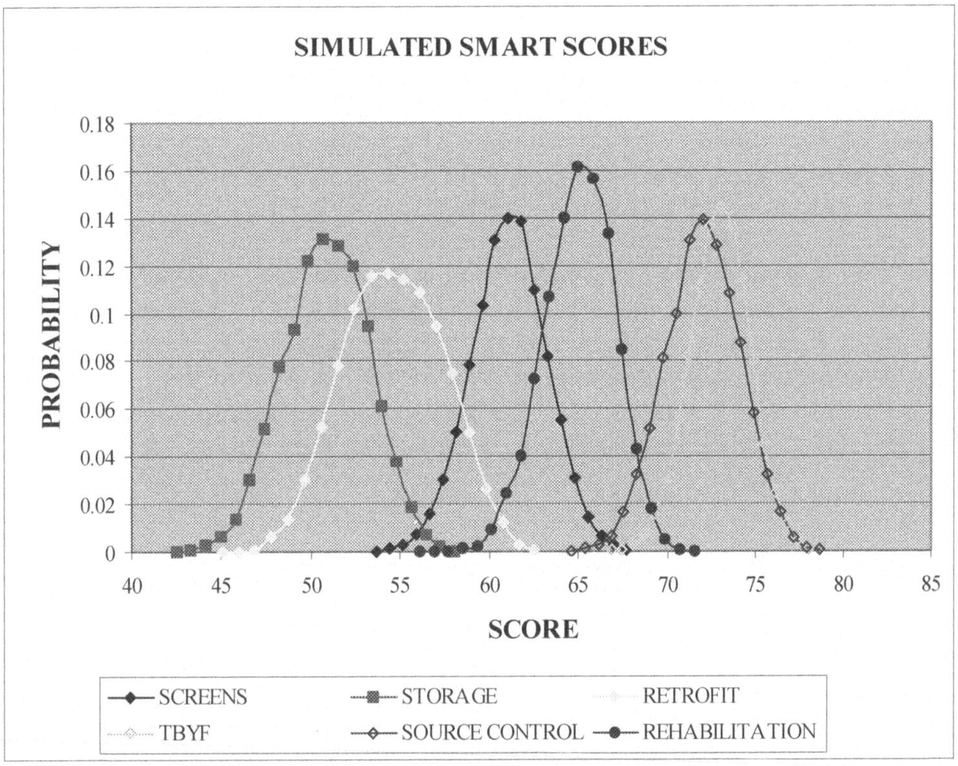

Figure D2.47: Simulated Smart Scores

D2.8.2 Making the Decision – the application of human knowledge and experience.

In order to simulate the final decision making stage, the results of the MCDA and the risk analysis were presented to a group of water industry experts with different professional backgrounds. The group were asked to review the data provided and to apply their judgement and experience to:

- to facilitate the inclusion of any intangible aspects not yet considered; and
- to reach a final decision.

Discussion

An initial discussion focussed on the identification of any intangible issues not yet included in the analysis. A number of issues arose; some already had been included in the selection of criteria. These were:

- That options selected will be constrained by the need for the water industry commissioner to buy in to any novel or alternative approach to tackling a problem.
- That flexibility in the selection of options will also be constrained by an expected budget spend which may have been set earlier in the decision making process. This

may in some cases limit option choices, but it was noted that this could have been included in the ratings of the economic criteria.

- There was some concern about selecting options where high levels on public participation were required. This is present within the decision making team even though this issue has been taken into account by the criteria. It is seen as one of the key overriding issues considered.
- Regulatory impediments are a problem when trying to select the best option. The group were not certain that it be possible to get agreement from regulators to remove a screen by implementing another option.
- In general the group identified a lack of confidence in selecting unproven methods, this issue exists in the decision makers minds as well.

Selection

Following a discussion consensus was reached that a combination of unproven methods, which were generally identified by the MCDA analysis as being more sustainable, and a complimentary proven method was appropriate. The public relations value of the TBYF and source control options were noted and deemed to be complementary as visit the houses to deliver rainwater barrels could be combined with TBYF education.

However, the final decision was to adopt a combination of TBYF and screen solutions with the screen giving a safety net if the TBYF is not successful. This was particularly important for this catchment which was a holiday resort. In such catchments, holidaymakers may not have the same knowledge and understanding of TBYF issues and therefore screens would still be required. However, it was noted that when screen replacement was required (20yrs), the success of TBYF could be established and the screen may not be necessary.

D2.9 POST-PROJECT MONITORING AND FEEDBACK

The post-project monitoring stage entails:

- **decision review** to verify that the selected option is that which contributes the most to enhancing sustainability.
- **process review** to refine the generic and specific criteria, the database and indicators (where used) and the SWARD processes.

D2.9.1 Decision Review

The Guidebook sets out the framework for this stage. This stage seeks to determine if the right decision was made.

Aims and scope of the appraisal

It is expected that appraisal would be undertaken some time after implementation of the preferred option. It is recommended that this should be at least one year in order to allow for an 'equilibrium' to be established in system performance.

The selected 'preferred' option was a combination of public awareness campaigns (TBYF) and screens. The screens are already installed in the catchment. TBYF campaigns have been operated and measurements of response made by interview and wastewater system monitoring, soon after the campaigns were run (Swain & Souter, 2002). Thus public behavioural change has already been promoted and response assessed.

The aims of the appraisal are to:

- Assess the robustness of the selected option in achieving the project objective (*to manage the number of sanitary waste items escaping to the environment via storm overflow discharges in the catchment in a way that is likely to be the most sustainable).*

- Assess the potential that other options if selected, may have led to greater sustainability. In this case study it is possible that an alternative combination of the original 6 basic options, not originally investigated, may have been more effective. In the original study only limited investigations of combinations were undertaken using one analytical method.

The scope of the appraisal may also address a wider perspective: -

- The original objective may be reviewed as additional factors (constraints and opportunities) may have become important since the initial study was conceived and implemented.
- A review of the options examined in the initial study and whether these were those most appropriate and whether they could now be added to.

Specify criteria to be evaluated

The criteria are the primary means by which relative sustainability between options has been evaluated. Thus they are the most important elements in the DSP. Post-Project appraisal should review the criteria -

- Test 1: Were the criteria, secondary criteria and indicators selected for the original evaluation fulfilled the specification for criteria selection and scope? Were they: -
    - Comprehensive
    - Applicable
    - Tractable
    - Transparent
    - Practicable
- Test 2: If the above were not fulfilled, then whether or not additional or revised criteria are required.

It is recommended that a review is made of the criteria using stakeholder groups once the selected option has been in operation for a year. The stakeholder groups should be both internal and external to the WSP. Stakeholder views as to the 'performance' of the development should be sought with the objective of assessing whether the originally selected criteria fulfilled the specification Tests 1 and 2 above and if changes need to be made.

Identify performance measures for the criteria

The criteria used in the original evaluation are shown in **Tables D2.4 –D2.7**. The indicators to be used to evaluate the criteria are also given. The performance measures are the indicators originally used but these should be reviewed at this stage. In the analysis, much of the information has been generated using simulation modelling to predict the performance of the various options. Following implementation of the preferred option (by construction or other means, such as public awareness campaigns) the actual performance (indicators) may be measurable. **Table D2.53** illustrates the indicators and where these would be expected to be measurable once the selected option has been implemented.

As it is the improvement of relative sustainability that is the overall objective, the performance measures should assist in confirming that the selected option is the best in this respect. Unfortunately, without implementation of each option, the assessment of relative performance of the options is not possible. Hence the objectives at this stage are to confirm that the predicted indicators in the initial DSP for the implemented option were correct. Measurements may also be conducted over a prolonged period to determine if the performance in relation to the criteria is deteriorating with respect to 'sustainability'.

Table D2.53 Indicators and potential for measurements following implementation of selected option

Criteria	Units	Change to enhance sustainability	How determined In initial analysis	Opportunities for actual measurement post implementation
T1 Sanitary Waste Escape	kg/yr	Reduce	Amounts of solids originating based on social surveys and checked with screen capture. Escape estimated by pro-rata volume splits for storm overflows using sewer flow simulation model.	1. Carry out TBYF surveys to verify inputs of SW (EV3). 2. Monitor flows to verify flow and volume splits at CSOs. 3. Undertake beach and marine surveys of SW.
T2 SW Transport	Qualitative	Reduce	This was assessed in terms of the propensity to be detained in the sewer network by qualitative means.	This could be assessed from the hydraulic model with depth and velocity thresholds*, or by CCTV in-sewer inspection.
T3 Risk of failure	Qualitative	Reduce	Qualitative assessment from expert discussion.	This could be determined with confidence only from long-term data over several years as to the overall performance.
T4 Flexibility and adaptability	Qualitative	Increase	Qualitative assessment from expert discussion.	As above, and only following the imposition of new constraints or introduction of new innovations.
E1. Capital cost	£	Reduce	Actual estimates from suppliers etc.	The real as implemented costs should be available.
E2. Operational cost	£/yr	Reduce	Estimated from suppliers and historical data	Collect data over the first year of implementation
E3. Maintenance cost	£/yr	Reduce	As above	As above
E4. Financial risk exposure	Qualitative	Reduce	Qualitative assessment from expert discussion.	Following implementation this may only be further assessed by comparison with other similar implemented schemes.
S1. Acceptability	%	Increase	Social survey	Follow-up social survey
S2 Perceived impact on the environment	%	Reduce	As above	As above
S3 Participation and responsibility	Qualitative	Increase	As above	TBYF survey or equivalent to assess actual public 'take-up' and involvement
EV1 Energy use	MJ	Reduce	LCA	Collect data over the first year of implementation for direct energy use. Energy in materials and recycle

				potential not readily measurable.
EV2 .1 Impact on air - CO_2	kg	Reduce	LCA	Collect data over the first year of implementation for direct emissions (if significant). Effects further up supply chain more difficult to determine.
EV2.2 Impact on air – Nox	kg	Reduce	LCA	As above
EV2.3 Impact on air – SO_4	kg	Reduce	LCA	As above
EV3 CSO Discharge	m^3/yr	Reduce	Computer model simulation	Flow monitoring over minimum of one year.

* Knowledge is limited about the transport processes for SW in sewers.

Using T1 as an example, actual in-sewer sampling and screen inlet monitoring at the WwTW has indicated that post-implementation TBYF campaigns have produced a measurable reduction of 65% in SW, and from public attitude surveys, a supposed reduction of 91%. However, the three 7–day monitoring periods for actual sanitary wastes were of short duration and may not be sufficiently representative to be confident that these reductions were genuine or sustained. Hence it is recommended that screenings weights be continually recorded at the WwTW from the dry and wet weather outlets in order to see trends in increases or decreases of SW inputs and consequent implications for indicator T1. Annual follow-up TBYF campaign refreshers should be undertaken until such time as there is confidence in public behavioural change and low levels of SW entering the system. This is essential if the preferred option of phasing out screens at the end of their current life is to be implemented.

Where additional criteria are identified that were not originally included in the DSP (Test 2), new secondary criteria/indicators will have to be defined and data collection will be required to define these.

Identify methodology for monitoring and data collection

At this stage the budget for monitoring will have been determined. This will reflect the priorities of the WSP. It is probable that monitoring will be undertaken for regulatory purposes only, rather than to verify that the option selection process was robust. Hence monitoring would be expected to be only for the factors in **Table D2.54** for this case study.

Table D2.54 Probable routine monitoring for SW case study

Monitoring	Measurements	Frequency
In-sewer flows (arriving at WwTW)	Flow rate, volumes and spill frequency at CSOs	Continuous
Limited inspections for SW escape (probably by local stakeholders)	Numbers of items	Semi-continuous
CCTV condition inspections	Visual information about in-sewer deposits	Once every 5-10 years
SW collected by screens and removed by road	Volume (m^3)	Continuous
Operational costs	£	Annual
Maintenance costs	£	Annual
Direct energy use	MJ	Annual

Some of these routine factors are appropriate for post-project assessment of the performance of the implemented option in terms of the sustainability criteria **in Table D2.53.** These can provide an initial review of the accuracy of the values for the indicators used when the decision was made.

The other criteria may be easily measurable with limited resources. For example, social surveys are usually relatively inexpensive. The main difficulties are in the 'supply-chain' information and database used for the atmospheric emission and LCA studies. Data for these analyses rely heavily on manufacturer provided information and standard baseline information for e.g. concrete manufacture. This is not amenable to direct measurement by the decision makers team, and decision makers should ensure that they are provided with as up to date information as possible about these factors.

Initiate study

This stage depends on whether the decision maker will implement a special study to investigate the robustness of the decision and performance of the implemented option or only wishes to utilise routine data as normally collected, as described in section D2.9.1. It is expected that the initial study would use routine data, possibly supplemented by limited and low cost additional information, such as the stakeholder perception of the success of the implemented option. Where new criteria have been identified these need to be quantified using additional data. It is recommended that prior to any measurements of the indicators for any new criteria, the way in which the criteria will be utilised in the post-implementation appraisal should be defined. This will require a review of the entire SWARD DSP used to make the original decision.

Acceptability of measured criteria

As it is likely that the decision review will utilise only routinely collected data, the initial appraisal may suggest the need for more detailed studies specifically to investigate each criterion. Acceptability of each criterion would be when the monitored information confirms that the indicator values used in the decision stage were accurate and within the bounds investigated in the sensitivity studies. The monitored information should also be used to refine the original selected option details. For example, where design and implementation has been based on particular operational conditions, these may require some modification to improve performance. Failure of criteria by demonstration that the original estimated or predicted indicators were inaccurate may also suggest that further criteria should be included in the appraisal. These may be additional criteria from the SWARD generic set, or entirely new criteria. It is recommended that ew criteria are introduced only after strict verification that they fulfil the specification in section D2.9.1 above and are not already included in any of the generic criteria.

D2.10 REFERENCES

Ashley R.M., Souter N., Dunkerley, J., Hendry S., Blackwood, D.J., Moir J. et al. (1999), Domestic sanitary waste disposal: Sustainability of alternative strategies, EPSRC Towards Sustainable Cities: Grant Reference GR/K62125, GR/K63375, GR/K63269.

CIRIA (2000). Sustainable urban drainage systems: design manual for England and Wales. Environment Agency report W230, CIRIA, London.

Foxon, TJ., Butler, D., Dawes, J., Hutchinson, D., Leach, M., Pearson, P. and Rose, D. (2000), An assessment of water demand management options from a systems approach, Journal of Chartered Institute of Water and Environmental Management 14, pp. 171-178.

InfoWorks (1999), Wallingford software.

Goedkoop, M. (1995), The Eco-indicator 95, final report. PRé consultants, The Netherlands.

PRé consultants B.V. (1997), SimaPro Single user database manual.

Swain, A and Souter, N. (2001/2), An assessment of the impact which promotional activities have on the volume, mass and speciation of sanitary items disposed of via the WC. Draft report for a Scottish Water Authority (2001). Draft report No. 2 dated January 2002.

Virtanen, V., Miettinen, P. and Junttila, V. (Eds.) (1995), Energy issues in Life Cycle Assessment. COMETT II UETP-EEE, Technical Research Centre of Finland.

Welsh Water & Friends of the Earth (1993), Odds and pods, March.

INDEX

IWA Publishing's authorised EU representative for General Product Safety Regulations is Diane D'Arras, 15 rue Duret, 75116 Paris, France, e-mail: safety@iwap.co.uk.

Printed and bound by CPI Group (UK) Ltd, Croydon, CR0 4YY
12/05/2026
02108881-0017